T0186333

Design and Operation of Activated Sludge Processes Using Respirometry

Alan F. Rozich
Anthony F. Gaudy, Jr.

LEWIS PUBLISHERS
Boca Raton   Ann Arbor   London   Tokyo

Library of Congress Cataloging-in-Publication Data

Rozich, Alan F.
    Design and operation of activated sludge processes using
respirometry / Alan F. Rozich and Anthony F. Gaudy, Jr.
        p.   cm.
    Includes bibliographical references (p.   ) and index
    1. Sewage—Purification—Activated sludge
process—Mathematical models. 2. Microbial
respiration—Mathematical models. I. Gaudy, Anthony F.
TD756.R69   1992
628.3'54—dc20   91-34375
ISBN 0-87371-449-0

LEWIS PUBLISHERS, INC.
121 South Main Street, Chelsea, Michigan 48118

Printed in the United States of America   3 4 5 6 7 8 9 0

Dedication:
To Gemma, Anton, Adrienne, and Libby

Alan F. Rozich received a BSCE with an emphasis in environmental engineering from Ohio State University in 1976 and an MS in environmental engineering from Ohio State in 1978. He worked for three years as a wastewater engineer for the city of Columbus, Ohio. Dr. Rozich received his Ph D in 1982 from the University of Delaware in environmental engineering. He worked on developing predictive models for activated sludge systems treating inhibitory or toxic wastes. Since 1983, Dr. Rozich has worked in various research, development, and consulting capacities. His efforts focused on developing improved techniques for designing and operating biological wastewater treatment systems with an emphasis on toxic, hazardous, and difficult-to-degrade wastes. Much of this work was directed toward utilizing respirometric methods as a means to calibrate predictive models for biological treatment systems. This predictive modeling technique has been applied to relatively diverse biological treatment situations as exemplified by a project involving the formulation of modeling strategies for predicting bioremediation rates at Prince William Sound, Alaska and one involving the formulation of a modeling approach for predicting landfill gas production rates. Dr. Rozich has presented and published over 50 technical papers and co-authored a book on biological waste treatment technology. He is a licensed professional engineer and also the holder of a process patent in waste treatment technology.

Anthony F. Gaudy, Jr. received a BSCE from the University of Massachusetts, his MSSE from the Massachusetts Institute of Technology, and a PhD from the University of Illinois, Urbana. Dr. Gaudy's career in the water pollution control field spans nearly 40 years and involves research, teaching, and consulting to industry and government. He has specialized in the biological aspects of pollution control because, early on, it appeared to him that this area comprised the most important and fundamental route to preserving the life support system. Staying on this track has enabled him and his 80 MS and PhD students to engage in what has amounted to a continuing comprehensive program of investigation into many aspects of the carbon-oxygen cycle and ways to achieve practical engineering control of this cycle. The work has sought to stress elucidation of fundamentals and, of more practical use, engineering methodology for optimal utilization of these fundamentals. Dr. Gaudy has authored more than 200 publications, and is H. Rodney Sharp Professor Emeritus, University of Delaware Department of Civil Engineering. This text brings together several of his long-term interests in modeling the microbial growth and respiration processes and their embodiment in a practical model to predict performance and provide operational control of the activated sludge process.

TABLE OF CONTENTS

PREFACE

Why write a book about respirometry and activated sludge?

This is an obvious question which the reader may ask. Indeed, why write this book. Many of the reasons have to do with the authors' experience concerning the research, development, and application of process control models for biological treatment. We realized that there was no one course which integrated the use of respirometry and the application of process control models for analyzing aerobic biological treatment systems. Our experience is that the use of process control models which are calibrated using respirometry represents a rapid, accurate, and cost-effective technique for generating process information for designing and operating aerobic biological treatment systems.

Our experience with field applications indicated that process models which are presented in this book do a reasonably good job of predicting the behavior of full-scale biological treatment systems. The authors also realized through feedback from consulting clients that the level of effort which was required to calibrate the model was a major issue which prevented its routine use and application. Calibration of the model primarily consists of determining the relationship between biomass growth rate and substrate or waste concentration. Methods such as the shake flask technique or the substrate utilization method were adequate for laboratory needs but unwieldly and resource-intensive for routine application. This led us to undertake a major research and development effort to develop a more cost-effective technique for calibrating the model.

Respirometry or the measurement of oxygen uptake rates of biomass has long been employed for many applications in the pollution control field. The basic premise relates to the simple fact that oxygen uptake rates provide a rapid indicator of microbial activity. In order to use respirometry for calibrating the model, one must be able to utilize the oxygen uptake data to compute cell growth or substrate utilization rates. A major effort which we performed showed that respirometric data can be employed to determine cell growth rates through transform of the data into equivalent growth or substrate utilization data. This is accomplished by utilizing equations which are derived using the principle of COD and energy balances for aerobic systems. Several research and field application efforts validated the use of respirometry as a means to obtain values of the biokinetic constants for model calibration. The model predictions for field reactors made using respirometric calibration accurately described the performance of field units. The key concept is that respirometry is used in these applications to calibrate a model which predicts the behavior of aerobic biological treatment systems.

The model which is presented in this book has its origins with work which was presented over 40 years ago. In the early 1950s, Monod and Novick and Szilard developed the theory of continuous culture. The significance of this work was to postulate that an engineering control such as reactor flow rate can be utilized to select the growth rate of a microbial culture. If the relationship between cell growth rate and substrate concentration is known, then one can obtain a desired effluent substrate concentration by using the appropriate engineering control such as flow rate. This concept is known as the "Theory of Continuous Culture."

During the 1960s, one of us (AFG) performed extensive work to show that the theory could be applied for predicting the results of heterogeneous microbial systems like those employed in activated sludge treatment processes. This work was necessary because the initial efforts of Monod, and others, focused on pure culture systems. Further developmental efforts started in the early 1980s extended the predictive modeling work for substrates which are inhibitory to microbial growth. Inhibitory kinetics are characteristic of many systems which must handle toxic or hazardous wastes or materials. In the mid-1980s, we began looking for kinetic techniques which could calibrate the model (define the relationship between growth rate and substrate concentration) quickly and cost-effectively. This led to the respirometric methods and the development and application effort for the technology which is described in this book. We now have this technology incorporated into a practice which we use routinely for engineering projects involving biological treatment systems.

How is this book set up?

The book contains seven chapters. The first two chapters provide a review of the fundamentals of modeling aerobic biological treatment systems. Chapter 3 gives the derivation of predictive equations for activated sludge systems while Chapter 4 compares and reconciles the modeling approach presented in this book with other methods. The methodology for using respirometry to obtain biokinetic constants is described in Chapter 5. Chapter 6 is important because it reviews the various environmental factors which can influence the values of the biokinetic constants which in turn impact process performance. It needs to be emphasized that the advent of respirometric techniques such as those described in the book enable environmental professionals to predict the impact of environmental conditions such as temperature, pH, and other factors on process performance much more quickly and accurately than that which one can realize with conventional approaches. Chapter 7 presents several case histories which describe the use of the respirometric technology for analyzing various design and operational situations.

Each chapter contains Introduction, Key Concept Summary, and References and Suggested Additional Reading sections. The use of an Introduction section is self-explanatory. The Key Concept Summary sections are designed to give the reader "bullet" summary versions of the important technical concepts which are presented in a particular chapter. The References and Suggested Additional Reading sections are given to provide those who are interested with more detail and background on the technical information presented in the text. The authors wish to emphasize that this book is not intended as a general textbook on biological treatment processes. Its purpose is to communicate information concerning a specific technical practice area which the authors developed and utilize. Consequently, only a minimum number of references are given in order to provide the necessary scientific and engineering back-up.

Who can benefit by using this book?

The main purpose of the book is to inform practitioners in the environmental field, especially those involved with biological treatment processes, about the utility of respirometrically calibrated models for analyzing biological systems. One interesting trend which we have observed is that clients who understand the technology suggest new ways to apply it for solving their problems or for enhancing operations. We liken this technology to a personal computer (e.g., a "Mac") in that once the fundamentals are understood, the number of applications is limited only by the imagination of the user.

At ERM, we have utilized this technology on numerous projects. Many times, we have been able to save clients both time and money. In the environmental business, time is often money, especially when there is need to fast-track a project in order to meet compliance deadlines. Projects have ranged from relatively basic treatability work to utilizing the respirometric approach to design thermophilic aerobic biological systems to treat a groundwater containing 200,000 mg/L COD.

What is the future of this technology?

As this book goes to publication, we are extending the boundaries and applications of the technology presented herein. We have completed projects which took the concepts for the respirometric calibration of activated sludge models and applied them to other areas. For example, work was performed that developed a modeling strategy for predicting bioremediation rates at oil contaminated beaches which resulted from the oil spill at Prince William Sound. Other work modified the aerobic modeling technique for use in predicting anaerobic methane production rates in landfills. Other applications which are in the works in our group include the development of standard protocols for determining the biodegradability of plastics and plastics substitutes and modifying the techniques presented herein for performing predictive modeling of composting (aerobic and anaerobic) systems.

Who else contributed to this work?

The authors wish to acknowledge some key individuals whose contribution and collaborative efforts were crucial in developing this technology. Mr. Richard J. Colvin has been working with us since 1983. He was instrumental on research and development and consulting projects which involved the development and application of respirometric methods. Dr. Elizabeth T. Gaudy ably served as a technical critic, proofreader, and coworker. She also ensured that our small consulting operation ran smoothly.

Messrs. Jerrold Wingeart and Paul H. ("Kip") Keenan, both of the City of Baltimore Public Works Department (Kip is now with Westinghouse), were instrumental in working with the authors to support the development and implementation of the technology. Their involvement made our efforts a professionally rewarding experience.

Finally, we wish to thank our technical reviewers for the time they took to examine and critique the manuscript. Mr. Donald Loftus of Star

Enterprise Refinery in Delaware City, Delaware and Dr. Paul J. Usino-wicz of ERM, Inc., Exton, Pennsylvania made valuable suggestions for this book. They have our gratitude.

Alan F. Rozich, Ph.D., P.E.
Anthony F. Gaudy, Jr., Ph.D., P.E.

LIST OF SYMBOLS

COD	chemical oxygen demand (mg COD/L)
COD_o	initial soluble COD for a batch test (mg COD/L)
COD_f	final soluble COD for a batch test (mg COD/L)
COD_t	soluble COD at time t (mg COD/L)
SCOD	soluble COD measured for a given waste sample (mg COD/L)
TCOD	total measured COD of a given sample (mg COD/L)
ΔCOD	soluble readily metabolizable COD of waste sample, COD_o–COD_f (mg COD/L)
ΔCOD_t	soluble readily metabolizable COD of waste sample over a given interval of time (mg COD/L)
ΔCOD_x	the change in the amount of COD incorporated into the cells (mg COD/L)
C.O.V.	coefficient of variation
D	dilution rate, influent flow rate, F, divided by aeration tank volume, V (h^{-1})
d	time in days
F	wastewater flow rate (mgd)
F_R	recycle flow rate (mgd)
h	time in hours
k_d	maintenance energy coefficient or specific decay rate (h^{-1})
K_i	inhibition constant for inhibitory wastes based on ΔCOD (mg COD/L)
K_s	saturation constant, substrate concentration, based on ΔCOD, at which specific growth rate is one-half the maximum rate (mg/L)
n	number of samples

$\Delta O_{2,t}$	cumulative O_2 uptake concentration at time t (mg O_2/L)
O_x	COD per mg of biomass
$\overline{O}_x$	average COD contained in a mg of biomass for a particular batch study
PST	primary settling tank
RAS	return activated sludge
S	readily metabolizable COD measured at time t (mg COD/L)
S_{avg}	average readily metabolizable COD concentration over a given time interval (mg COD/L)
S_e	readily metabolizable effluent COD leaving plant (mg COD/L)
S_f	readily metabolizable final COD measured in a batch system (mg COD/L)
S_i	readily metabolizable influent COD entering plant (mg COD/L)
S_o	readily metabolizable initial COD measured in a batch system (mg COD/L)
S_R	readily metabolizable recycle COD measured within the plant (mg COD/L)
S.D.	standard deviation
t	time (h or d)
$\overline{t}$	hydraulic detention time (h)
V	reactor volume (million gallons)
X_f	largest concentration of biological solids measured during the substrate removal phase of a batch test (mg TSS/L)
X	concentration of biological solids (mg TSS/L or mg VSS/L)
X_o	initial concentration of biological solids in a batch test (mg TSS/L)
X_R	concentration of biological solids in the recycle flow to the reactor (mg TSS/L)
X_t	concentration of biological solids at time t (mg TSS/L)
$X_{t\,avg}$	average concentration of biological solids over a given time interval (mg TSS/L)
ΔX_t	change in the concentration of biological solids over a given time interval (mg TSS/L)

Y	cell or sludge yield; mg biomass produced per mg COD metabolized (mg/mg). In a batch system the observed yield is the true yield. In a recycle system, the true sludge yield is calculated from a maintenance plot.
α (alpha)	recycle flow ratio (F_R/F)
μ	specific growth rate (h^{-1})
μ_{max}	maximum specific growth rate (h^{-1})

1 FUNDAMENTALS OF BIOKINETICS FOR ACTIVATED SLUDGE SYSTEMS

INTRODUCTION

The principles of the activated sludge process are really not very difficult to understand. In fact, approximately four decades ago it was thought by some that we already knew all we were ever going to know about the process and that further study was pointless. After all, the saying went, "the bugs eat the waste"; what else did we really need to know? Today, however, some people take the reverse attitude. They feel that the activated sludge process has failed to serve the engineering profession because it is too complex to be understood clearly enough, and thus controlled precisely enough, to deliver the type of effluent quality demanded by today's environment.

Both the oversimplified attitude and the overcomplex attitude give erroneous impressions about the process; the truth lies somewhere between the extreme views. One of the authors of this book has enjoyed a career spanning four decades (mentioned above) in the attempt to uncover and clarify some of the complexities to simplify engineering of the process and promote its goal of wastewater purification. In recent times, the activated sludge process has been increasingly called on to treat not only readily metabolizable wastes but also those containing toxicants. There is much evidence that toxic wastes can be successfully treated provided one adheres to certain fundamental principles. It is the aim of this text to set down such principles in a quantitative and practical fashion and to array them in a useful engineering methodology readily

1

applicable not only to process design but to process operation as well. It should be apparent that, to possess engineering value, design models must also possess the analytical properties needed for guiding operations as well as initial design of the process.

The goal of this chapter is to describe the purification process in terms of microbial growth, which is the mechanistic basis for the biological treatment of wastewaters. Following a brief description of the process, the parameters needed to quantify growth, purification, and oxygen uptake (i.e., exertion of biochemical oxygen demand, BOD) are given. The necessary biokinetic parameters are established, and relationships between growth, purification (i.e., substrate or waste removal), and respiration (i.e., O_2 uptake) are presented. After studying this chapter, the reader should have a solid grasp of the quantitative concepts required for understanding the development and calibration of models that describe the microbial purification of wastewaters containing both toxic and non-toxic carbon sources.

SOME GENERAL PRINCIPLES

Perhaps the appropriate scientific phraseology for the idea "the bugs eat the wastes" is as follows: "The heterogeneous microbial population utilizes waste materials to obtain energy and to grow." They use the waste material in the same manner in which we use food materials for energy and for growth. In order for the microbes to accomplish this, the waste molecules must be soluble or made soluble so that the process is not hampered simply by hindered access of the food molecules to the microbial cells. All the primary feeders (bacteria) in the population use soluble food. The secondary feeders are larger microbes, e.g., protozoa, which ingest particulate food, mostly bacteria that have grown on the waste molecules. The primary and secondary feeders, along with whatever particulate matter is contained in the waste or passes into the aeration tank, are collectively termed "activated sludge."

From a chemical standpoint, this sludge consists mainly of several classes of biochemical compounds that are characteristic of all living matter (carbohydrates, proteins, lipids, and nucleic acids). Synthesized by all living systems, these biochemical compounds contain mainly carbon, which the cells obtain from the compounds in the waste. Some of these compounds also contain nitrogen, phosphorus, and many other elements in lesser amounts. Naturally, all of the elements needed to synthesize the biochemical compounds must be present in the wastewater. When the needed compounds are out of balance, those in short supply must be added. For example, nitrogen and phosphorus are often necessary additions to some industrial wastewaters, which usually contain such a large supply of carbonaceous material that growth could not occur in a balanced manner without supplementation of nitrogen and phosphorus. The aim in the treatment process is to ensure that carbon is the growth-limiting nutrient and that growth is balanced.

Usually, the chemical composition of the sludge is not of major interest in the activated sludge process, and microbial (sludge) growth is usually assessed as an increase in the mass of sludge in the system regardless of its composition. The reader should realize that balanced growth is not necessary for growth to occur (that is, growth as measured as an increase in the mass of material). The mechanism of oxidative assimilation, an important mechanism that sometimes occurs in treatment systems, is an example of unbalanced growth in which large amounts of nonnitrogenous carbon compounds such as carbohydrates and lipids may be synthesized with little or no production of protein or nucleic acids. Such unbalanced growth can at times provide for the removal of large amounts of substrate. However, it should be remembered that in order to maintain the substrate removal capability of an activated sludge, the biomass that is eventually recycled to the aeration tank should have the opportunity to synthesize nucleic acid and protein.

Oxygen is supplied to an activated sludge reactor in order to provide the microorganisms with the means to oxidize a portion of the organic compounds in the waste material. A portion of the energy released during the oxidation process is converted into chemical energy, which permits the organisms to use the remaining portion of the organic carbon as building blocks to synthesize the particular compounds (carbohydrates, proteins, lipids, and nucleic acids) needed for their structure and function. Thus, a wastewater exerts an aerobic biochemical oxygen demand (BOD). The amount of oxygen used during the process of obtaining energy and extracting organic carbon from the wastewater for growth is a measure of the biochemical oxygen demand exerted during the metabolic purification of the wastewater. Thus, the exertion of BOD and the growth of microorganisms, that is, sludge production, go on concurrently and are interrelated processes. The sum or total effect of these interrelated processes is the removal of the carbonaceous material from the waste water, i.e., purification of the waste. After the organic carbon has been either oxidized or taken up by the microorganisms in the growth process, oxygen uptake will still occur because the newly synthesized population begins to undergo autodigestion or endogenous respiration.

While the biochemical and mechanistic processes are of much interest, most engineers are concerned more with quantitative description of the rates of occurrence of these phenomena and the completeness of the process. This important descriptive exercise can be aided by the study of Figure 1.1. Let us assume (1) that the batch reactor shown in the figure has been loaded with the waste containing amounts of organic carbon compounds, nitrogen compounds, phosphorus compounds, etc., so that balanced growth occurs, and (2) that the reactor has been seeded with a small amount of activated sludge that has been thoroughly acclimated to the carbon compounds contained in the waste. Aeration has begun and waste constituents in solution are measured, in this case as chemical oxygen demand (COD). Sludge concentration (X), and O_2 uptake are also measured over time. In the figure, COD is indicated as a measure of the organic matter present in solution. It can be seen that the elimination of COD proceeds to some lower limit, i.e., there is a residual COD. Since the biomass seed was well acclimated and conditions of pH, temperature, nutrition, air supply, and mixing were optimal, the difference between the initial and the lowest COD (the end of substrate removal phase) is termed the ΔCOD and the numerical value of the ΔCOD is a measure of the strength of the waste. This term shall be designated as S_i in this text (see Equation 1.1).

$$S_i = \Delta COD = COD_i - COD_e \qquad (1.1)$$

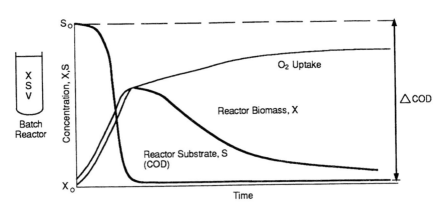

Figure 1.1.   **Course of substrate removal, biomass growth and decay, and oxygen utilization in a once-fed batch reactor under aerobic conditions (from Gaudy and Gaudy, 1988).**

At the end of the substrate removal or purification phase the ΔCOD has been either oxidized (i.e., registers as accumulated O_2 uptake) or channeled into new biomass. Beyond this phase, called the autodigestive phase, oxygen uptake increases but biomass decreases. In this phase some of the primary feeders, bacteria, are used as food by the predator population, largely the protozoa. The oxygen that is used in this phase consists primarily of O_2 used by the protozoa in obtaining energy when feeding on the bacteria and O_2 used by the remaining bacteria, which continue to respire endogenously.

It is of much interest to note here that the 5-day BOD, which is measured as a single point on the O_2 uptake curve, usually occurs long after the substrate removal phase, i.e., well into the autodigestive phase. Eventually O_2 uptake becomes asymptotic to some upper limit, the ultimate carbonaceous BOD, L_o. This value is always lower than the ΔCOD (both values are measured in terms of oxygen). Further exposition of the relationship between ΔCOD and L_o can be found in the literature (Gaudy and Gaudy 1988).

SOME IMPORTANT QUANTITATIVE CONCEPTS

In developing quantitative concepts, the substrate will be designated S. S in practical terms will be considered to be COD or more correctly, ΔCOD, but the reader should realize that other analyses could be used as a measure of substrate; for example, total organic carbon (TOC) or ΔTOC. Biomass or activated sludge will be designated X and O_2 uptake simply O_2.

Cell or Sludge Yield

One of the most practical quantities of interest is the cell or sludge yield, Y_t. This factor is the ratio of the amount of sludge, or biomass, produced per unit of substrate removed when the cells are growing rapidly. If one takes the biomass readings along the appropriate location on the curve of Figure 1.1, the cell yield is given by Equation 1.2.

$$Y_t = \Delta X / \Delta S = (X_t - X_o)/(S_o - S_t) =$$
$$(X_t - X_o)/(COD_o - COD_t) \qquad (1.2)$$

At any point during the substrate removal phase shown in Figure 1.1, Y_t remains constant. Of course, a fairly low value will be measured if the measurement is taken at some point in the autodigestive phase, and erratic values may be obtained if the measurement is made too early in the growth phase. We could measure the cell yield well into the endogenous or autodigestive phases, but the value one obtains should not be confused with Y_t, which is often called the true or maximum cell yield, i.e., the cell yield unaffected by autodigestion. When the cell yield value can be shown to have been affected by autodigestion, it is designated by the term Y_o. This is a very important parameter and is dealt with later when we discuss the production of excess sludge in continuous flow systems.

Growth Rate and Decay Rate

The growth rate of microbial mass dX/dt or $\Delta X/\Delta t$ is always expressed as the specific growth rate, μ (i.e., the rate of growth per average unit of biomass or cells extant during the time interval dt or Δt) and is given by Equation 1.3.

$$\mu = (1/X)(dX/dt) \tag{1.3}$$

Figure 1.2 shows various phases in the microbial growth cycle, which observers have used to describe the growth and decay of microorganisms. The phases that are of most interest are the logarithmic, increasing phase in growth and the autodigestive phase (mainly the decelerating portion) because in these phases the specific rates μ for growth and specific decay rate k_d for autodigestion are essentially constant. This offers a convenient manner of quantitative kinetic evaluation as described below.

The existence of exponential growth in a batch system is easily tested by plotting the values of X or some marker for X such as optical density (turbidity) against time on a semilogarithmic scale (see Figure 1.3). Quite simply, the exponential phase is the straight line portion. The slope, μ, is constant. During this phase, X at any time can be predicted by integrating Equation 1.3 as shown in Equation 1.4.

$$X_t = X_o \exp(\mu t) \tag{1.4}$$

The numerical value of μ for this system can be evaluated from the experimental data using Equation 1.5.

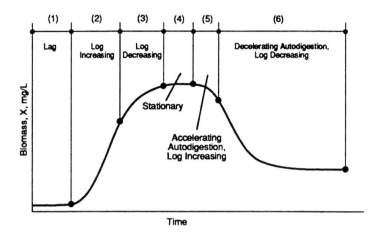

Figure 1.2. Distinguishable portions of biomass growth and decay curve (from Gaudy and Gaudy, 1988).

$$\mu = \ln(X_t/X_o)/(t - t_o) \qquad (1.5)$$

If one uses as the time interval the time it takes for a doubling of X, i.e., the doubling time t_d, the specific growth rate μ is given by Equation 1.6.

$$\mu = \ln2/t_d = 0.693/t_d \qquad (1.6)$$

A plot such as the one shown in Figure 1.3 is extremely useful since it provides a simple way to determine the existence and the extent of the logarithmic phase and also shows the beginning of the declining phase of growth. If there is a lag phase, a plot of the curve also shows when it ends and the logarithmic phase begins. This is very important in handling field data because the numerical values one obtains for μ must be obtained using only data points within the exponential phase. The more data points one obtains, the more accurate and useful is the value of μ obtained.

Similarly, the autodigestive phase is often characterized by a constant specific rate of decline in biomass. Figure 1.4 depicts the logarithmically decreasing autodigestion of a biomass, described by Equations 1.7, 1.8, 1.9, and 1.10. In these equations, $t_{0.5}$ is the time required for the population to decrease by one half.

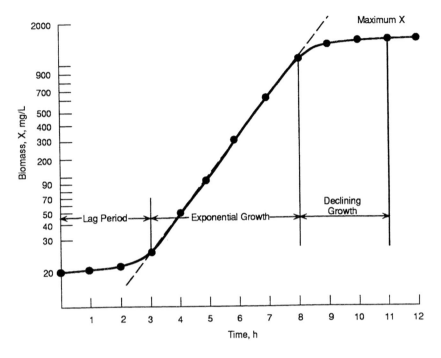

Figure 1.3. **Semilogarithmic plot of a biomass growth curve showing extent of exponential or logarithmic growth (from Gaudy and Gaudy, 1988).**

$$dX/dt = -k'X \qquad (1.7)$$

$$X_t = X_o exp(-k't) \qquad (1.8)$$

$$k' = ln(X_o/X_t)/t \qquad (1.9)$$

$$k' = ln2/t_{0.5} = 0.693/t_{0.5} \qquad (1.10)$$

Relationship Between μ and S (Noninhibitory Waste)

As we shall see, the specific growth rate μ is a vital system characteristic. It is affected by many physical and chemical environmental factors, such as temperature, pH, chemical inhibitors, etc. These will be discussed in more detail in Chapter 6. However, the factor which makes this

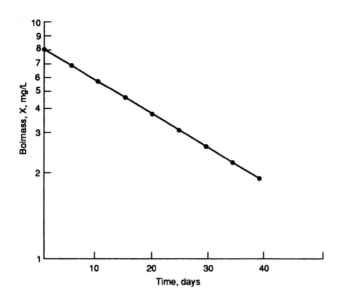

Figure 1.4. Decelerating autodigestion (logarithmically decreasing). Phase 6 of Figure 1.2 (from Gaudy and Gaudy, 1988).

parameter so unique in the description of the behavior of microorganisms toward a specific waste is the fact that its numerical values are governed mainly by the type of substrate and cells present. This fact makes μ a powerful tool for characterizing waste/activated sludge systems and predicting effluent quality. Moreover, μ is affected by the concentration of the waste carbon source or substrate. This fact is discussed below in developing models relating μ to the concentration of the waste. The specific growth rate at which a system is run is by far the most important factor in determining both the design of the wastewater treatment facilities and the amount of excess sludge to be handled.

If one sets up several experiments like the one shown in Figure 1.3, results such as shown in Figure 1.5 (part a, b, or c) are generally obtained. The initial concentration of wastewater is given in parentheses, and it is seen that the slope, i.e., μ, is higher for the higher concentrations of substrate. A plot of μ vs. S (see the lower graphs) shows that μ approaches some upper value as S is increased. The difference between systems a and b is that in b a higher concentration of S is required to make μ approach its maximum value than in a. Examination of the graph

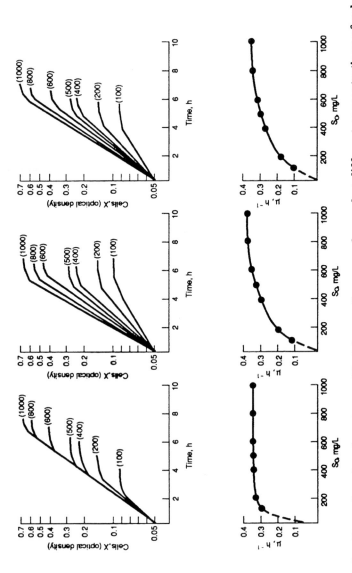

Figure 1.5.   Examples of various types of batch growth curves observed at different concentrations of substrate. (Top) Growth curves at initial substrate concentrations shown in parentheses. (Bottom) Specific growth rate versus initial substrate concentration (from Gaudy and Gaudy, 1988).

in part c shows that there is really no significant straight line portion (i.e., essentially no exponential phase at the concentrations shown). This is not the general case, and even when this happens, the specific growth rate varies according to the initial substrate concentration as well as the changes in S as the organisms grow. A more detailed discussion of these types of data sets can be found elsewhere (Gaudy and Gaudy 1988) but the main point to be remembered here is that when one analyzes the data to determine an equation describing the lower graphs, it is nearly always found that Equation 1.11 provides a fairly accurate fit of the experimental data. Equation 1.11 describes a rectangular hyperbola. The symbol μ_{max} designates the upper or maximum value of μ regardless of how high S_o is made and the symbol K_s is a term related to the flatness or sharpness of the curve as μ approaches μ_{max}.

$$\mu = \frac{\mu_{max}S}{K_s + S} \tag{1.11}$$

One can see that the value of K_s is numerically equal to the concentration of S which makes μ equal to one-half of μ_{max}. This equation is very familiar in the field of biological kinetics; it is called the Monod equation after Jacques Monod, who first demonstrated the fit of μ to S in accord with the rectangular hyperbola.

Figure 1.6 provides a numerical example of data description using the equation. The lower portion shows a linear form of the equation and the relationships needed to determine the numerical values of K_s and μ_{max} for a cell/waste system. Other methods for obtaining numerical values from experimental data are given in Chapter 5.

At times engineers may doubt, when dealing with essentially nondefined sewage and entirely heterogeneous microbial populations, that the relationship between values of μ at different concentrations of S would follow according to the Monod equation. However, Figure 1.7 should help allay any such doubts. This figure shows a plot of μ values obtained at varying initial COD concentrations for the soluble portion of a municipal sewage. The higher COD concentrations were obtained by concentrating the municipal sewage.

Relationship Between μ and S (Inhibitory Waste)

Often, wastewater contains some substances that inhibit growth as well as those that stimulate growth. Sometimes, the same waste constitu-

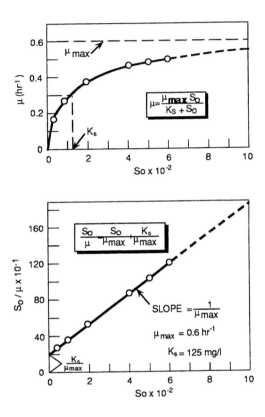

Figure 1.6.   **Hyperbolic plot of the relationship between specific growth rate, μ, and initial substrate concentration, S_o (upper graph), and a straight line plot, S_o/μ vs. S_o of the same data (lower graph).**

ent used for growth stimulation can also inhibit or retard growth. As for noninhibitory wastes, the specific growth rate depends on waste concentration. For such wastes, the most commonly demonstrated effect of S on μ is one in which μ increases with S up to some concentration beyond which further increase in S serves only to decrease μ. As with nontoxic wastes, there are several mathematical expressions that have been fitted to plots of μ vs. S, but by far the most commonly observed relationship is the one expressed by Equation 1.12. We will refer to it as the Haldane equation because of its similarity in form to the relation found by Haldane to describe substrate inhibition in some enzyme systems.

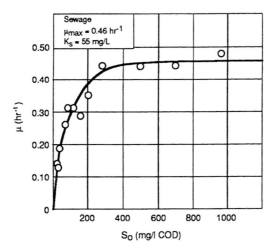

Figure 1.7. Hyperbolic plot of the relationship between specific growth rate, μ, and initial substrate concentration, S or S_o, for a heterogeneous microbial population of sewage origin growing on a concentrate prepared from the soluble portion of municipal sewage.

$$\mu = \frac{\mu_{max}S}{S + K_s + S^2/K_i} \qquad (1.12)$$

It is emphasized that both Equations 1.11 and 1.12 are selected for modeling of wastewater processes purely on the practical basis that they are found to provide rather good fits to experimental results obtained in systems important to environmental engineers. It is not necessary and is probably dangerous to assign general theoretical importance to these equations by bridging between descriptive formulations for enzyme kinetics and those for microbial growth, other than to realize that microbial growth rate is governed at the molecular level by enzymes. One should realize, however, that factors governing one enzyme reaction could not necessarily be expected to govern the overall kinetics of growth when one considers the complicated array of enzymes embodied in a living cell. Equations 1.11 and 1.12 are justifiably used in modeling because (1) they can be shown to fit real wastewater data in the field and (2) the numerical values of the biokinetic constants they employ have real quantitative value in predicting activated sludge treatment performance.

Examination of these equations shows that there are two biokinetic constants in Equation 1.11 and an additional one in Equation 1.12. The constant K_i is termed the inhibitory constant and it expresses the inhibitory nature of the waste to the growing cells. If its numerical value is very large, then inhibition is minimized and Equation 1.12 is the same as Equation 1.11.

Figure 1.8 compares the type of plot rendered by Equations 1.11 and 1.12. The difference in the type of behavior of μ with increasing S_o is apparent. Increasing concentrations of toxic substrates inhibit (depress) μ. The μ value does not approach some maximum μ not affected by further increase in S; rather, it goes through some maximum value at a specific concentration of S. This peak in μ is extremely significant in assessing the stability and resilience of systems growing on inhibitory wastes. It is designated the critical specific growth rate, μ^*. Its numerical value and the substrate concentration at which it occurs, S^*, are easily determined. Setting the first derivative of Equation 1.12 to zero and solving for S and μ yields Equations 1.13 and 1.14.

$$\mu^* = \frac{\mu_{max}}{1 + 2\sqrt{K_s/K_i}} \qquad (1.13)$$

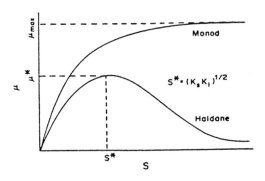

Figure 1.8. **Relationship between μ and S according to noninhibitory (Monod) and inhibitory (Haldane) equations for microbial growth on substrate. μ^* is the peak, or highest possible, specific growth rate on the inhibitory substrate and S^* is the substrate concentration at which $\mu = \mu^*$. Higher substrate concentrations cause a decrease in growth rate (from Rozich, Gaudy, and D'Adamo, 1985)**

$$S^* = \sqrt{K_s K_i} \qquad (1.14)$$

These equations yield numerical values of substrate and specific growth rate beyond which activated sludge systems treating the specific toxic wastewater in question cannot grow and will fail; that is, the biomass will wash out of the reactor and effluent quality will totally deteriorate. How these values are employed in operation and design is detailed in Chapter 3.

Nature of the Growth Curve

Analysis of the growth curves from which μ is obtained for the various values of S_0 deserves special comment. As with nontoxic substrates, exponential growth (straight line portion of a semilogarithmic plot of X vs. t) is observed, but growth is generally much slower at all S_0 concentrations than for the nontoxic substrate and is extremely slow at high S_0 values. In addition, at high values of S_0, the general shape of the growth curve after attainment of the exponential phase is distinctly different than for nontoxic substrates. After the exponential phase, μ actually increases, because as the cells grow, the substrate concentration is reduced to a less toxic value, permitting a higher specific growth rate.

Figure 1.9 shows growth curves for heterogeneous populations of sewage origin on an inhibitory substrate, phenol. Part *a* of the figure shows growth rates obtained at S_0 concentrations from 50 to 500 mg/L phenol COD, and in part *b*, concentrations range from 700 to 900 mg/L phenol COD. The graphs show that all the growth data in this experiment were obtained at S_0 concentrations above S^*. Note that the growth rates decreased for increasing S_0 values. It can also be noted that the curves for 500 to 1000 mg/L show increased slopes after the exponential phase. As noted above, this is only observed for toxic or inhibitory substrates when the system is initiated at S_0 values significantly in excess of S_0^*. In the case shown, S^* was 30 mg/L phenol COD. When initial substrate concentrations below S^* are employed, the general shape of the growth curve is the same as that for a nontoxic substrate. The difference in the shape of the growth curves on either side of S^* can also be demonstrated by a computational simulation of growth using the Haldane equation. Such a comparison is made in Figure 1.10 for S_0 concentrations ranging from $0.5\ S^*$ to $10\ S^*$. It is seen that at S^* and below, the growth curve is that for a typical noninhibitory substrate, i.e., an exponential or apparently exponential phase followed by a decreasing rate; whereas growth at

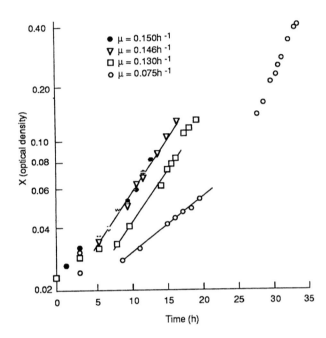

Figure 1.9a.   Growth on phenol at initial concentrations of 50 (●), 100 (▽), 200 (□), and 500 (○) mg/L. Lines show exponential phase (from Rozich, Gaudy, and D'Adamo, 1985)

S_o concentrations above S^* is typical of that of an inhibitory substrate concentration, i.e., exponential followed by increasing rate and finally decreasing rate as the carbon source is exhausted. It is important to be aware of these differences in behavior of the growth curves when developing and analyzing growth data from which numerical values of the biokinetic constants are to be determined.

Relations Between μ and S

Monod, Nontoxic Wastes

This relationship is defined by the specific growth rate μ, which increases with increased S_o values at a decreasing rate of increase, i.e., it follows the form of a rectangular hyperbola. Equation 1.11 is an empiri-

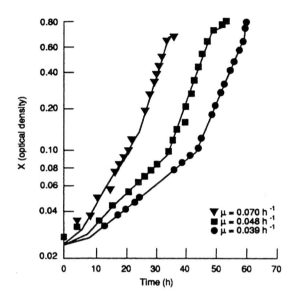

Figure 1.9b.   Growth on phenol at intial concentrations of 700 (▼), 900 (■), and 1000 (●) mg/L. Lines show exponential phase (from Rozich, Gaudy, and D'Adamo, 1985).

cal one based on observations of experimental results. The equation holds for heterogeneous populations and mixed carbon sources as well as for pure cultures and sole carbon sources. The biokinetic constants μ_{max} and K_s are dependent upon temperature, pH, etc., as are any chemical constants, but for given conditions or a range of conditions μ_{max} and K_s are characteristic of the waste and acclimated biomass growing on it.

Haldane, Toxic Wastes

This relationship is given in Equation 1.12. It is similar to the Monod equation but the last term of the denominator accounts for decreasing μ with increasing substrate concentration beyond a critical value of S, i.e., S*, which is given by Equation 1.14. The critical specific growth rate μ^* that occurs at this substrate concentration is defined by Equation 1.13. S* and μ^* are particularly important for design and control of bioreactors in which inhibitory wastes are to be treated because beyond this

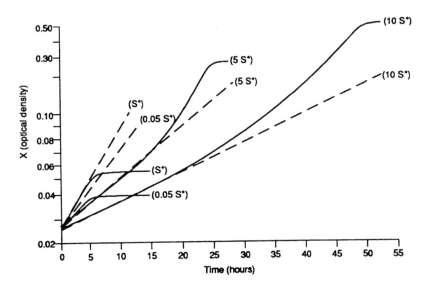

Figure 1.10. Dependence on initial concentration of the shape of batch growth curves according to the Haldane relationship between μ and S. $\mu_{max} = 0.34$ hr^{-1}, $K_s = 32$ mg/L, $K_i = 67$ mg/L, $Y_t = 0.00085$ O.D. units/mg substrate/L, $S^* = 69$ mg/L, and initial biomass concentration = 0.025 O.D. units. Dashed lines represent exponential growth (from D'Adamo, Rozich, and Gaudy, 1984).

substrate concentration and specific growth rate the system cannot exist for all practical purposes in most applications.

The Haldane equation becomes the Monod equation when a system exhibits very high values of K_i. Thus, there is a gradation from toxic to nontoxic wastes. One determines whether a waste exhibits a toxic or a nontoxic reaction by observing the nature of the experimental plots of μ vs. S in accord with Figure 1.8. If it is found that μ decreases after peaking at some value of S, then the system exhibits an inhibitory nature. Such an observation is important because it determines which type of equation should be used to design and operate the system. Note the values of μ^* and/or S^* provide a basis for assessing the severity of the toxic condition of the system of cells and waste or substrate.

Relation Between S, X, and O_2

There is a relationship between substrate removal, i.e., purification, growth of biomass, and oxidation (O_2 uptake) implicit in Figure 1.1. Figure 1.1 tells us that the COD being removed from solution is accounted for totally by biological oxidation and by growth of biomass. If the waste components were subject to rapid stripping or were very easily oxidized by oxygen this condition would not hold. Whether the waste is strippable or chemically oxidizable simply by aeration is easily checked; in most cases, chemical oxidation and stripping do not account for substrate removal. Thus the split of substrate removal between respiration (oxidation) and synthesis (growth) provides a good basis for making mass and energy balances on this aerobic biological system, and a useful methodology to check results and to use respirometry to obtain needed growth data. At various points along the COD removal curve of Figure 1.1 the amount of COD removed must be accounted for as the summation of that used for synthesis (i.e., the amount appearing as biomass) and the amount that has been oxidized (i.e., the amount that can be accounted for as oxygen uptake). These values can be easily measured and inserted in Equation 1.15.

$$\Delta COD_t \cong \Delta O_2 + \Delta X_t \tag{1.15}$$

The equation states that the amount of COD removed at any time is equal to the sum of the accumulated O_2 uptake to that time and the amount of cells produced. However, one must first express the amount of cells in terms consistent with the other terms. We already have COD expressed in terms of oxygen and the ΔX term can be converted to equivalent O_2 by determining the COD of the cells. We can do this either by direct COD analysis or by determining the unit COD per unit of biomass, O_x. This ratio can change after the substrate removal phase. The COD of growing cells is often somewhat higher than that for autodigesting cells. During the growth phase this value remains fairly constant, similar to the cell yield, Y_t. Numerical values of O_x should be determined for the wastewater system under study but for purposes of preliminary analysis, the O_x obtained from Dr. Porges' empirical formula for activated sludge may be used (Porges, Jascewicz, and Hoover, 1956).

$$C_5H_7NO_2 + 5O_2 \rightarrow 5CO_2 + 2H_2O + NH_3 \tag{1.16}$$
$$O_x = (5)(32)/113 = 1.42 \text{ mg } O_2/\text{mg biomass}$$

Equation 1.15 may now be written correctly as:

$$\Delta COD_t = \Delta O_{2,t} + \Delta X_t \cdot O_x \qquad (1.17)$$

Equation 1.17 can be used to check the accuracy of test data. It is essentially an energy balance. This relationship between O_2 uptake, growth, and purification can also be used in another important way. Since ΔX_t is related to ΔCOD_t by the cell yield, Y_t, Equation 1.17 can be modified to Equation 1.18.

$$\Delta COD_t = Y_t \cdot O_x \cdot \Delta COD_t + \Delta O_{2,t} \qquad (1.18)$$

ΔCOD_t is also defined by Equation 1.19.

$$\Delta COD_t = (X_t - X_o)/Y_t \qquad (1.19)$$

Equation 1.19 is substituted into Equation 1.18 to produce Equation 1.20.

$$X_t = X_o + \Delta O_{2,t}/(1/Y_t - O_x) \qquad (1.20)$$

With Equation 1.20 we can obtain the growth curve from the O_2 uptake curve, provided we know the values of Y_t and O_x, both of which are easily determined. This equation is extremely important because it permits one to make use of those parameters to obtain growth curves needed to evaluate the biokinetic constants μ_{max}, K_s, and K_i, which are employed in activated sludge models. In contrast to direct measurement of X_t by optical density or gravimetric means, both of which methods require much labor and expense, respirometric readings of accumulated oxygen uptake can be obtained rapidly and automatically and converted to X_t values using equation 1–20. Thus the copious amounts of data needed to define growth curves are easily obtained. The COD remaining in solution at any time can also be determined through Equation 1.21.

$$COD_t = COD_o - \Delta O_{2,t}/(1 - O_x Y_t) \qquad (1.21)$$

It can be seen that the denominator of the last term of Equation 1.20 combines two system constants and may be represented by a single term, R.

$$R = 1/Y_t - O_x \qquad (1.22)$$

The term R is defined as the respirometric ratio and has definite physical significance; it is the milligrams O_2 consumed per milligram of biomass produced. Thus the numerical value of R can also be depicted using Equation 1.23.

$$R = \Delta O_{2,t}/(X_t - X_o) \tag{1.23}$$

One can determine R using both Equations 1.22 and 1.23 as a check on the accuracy of the data.

Equations 1.20 and 1.21 can now be written as given below.

$$X_t = X_o + \Delta O_{2,t}/R \tag{1.24}$$

$$COD_t = COD_o - \Delta O_{2,t}/(R \cdot Y_t) \tag{1.25}$$

After determining the respirometric ratio, R, one can very easily produce the growth and COD removal curves from the respirometric data. (Gaudy et al. 1988, 1989, 1990). It should be emphasized that the equation pertains to the substrate removal phase even though the energy balance principle upon which these equations were derived is valid in both the substrate removal and autodigestive phases.

Equations 1.24 and 1.25 provide the engineer with a powerful tool for obtaining growth and substrate removal rate data which are used to determine the numerical values of biokinetic parameters for biological treatment systems. These values are in turn used with the reactor model to be presented in Chapter 3. However, before proceeding to the description of engineering models for activated sludge reactors, it is most desirable to present some basic kinetic reactor engineering principles of continuous growth, which when combined with the biokinetic principles just delineated form the basis for the engineering approach to be presented later.

KEY CONCEPT SUMMARY

It may seem an oversimplification to proclaim that since the beginning of aerobic biological treatment of wastewaters (especially by the activated sludge process) the status of its conceptual understanding can be condensed to the simple definitions and relationships represented in the 25 equations thus far discussed. While space considerations and the desire to simplify have necessitated avoidance of some important details and possible refinements of the concepts, the fact remains that, through continued research, understanding and use of these concepts are just coming into engineering practice in the field and the material presented thus far can be truly represented as "state-of-the-art."

Since understanding of the foregoing terms and relationships is such a vital foundation for practical application of material that follows in this text, it is important to summarize the key terms and related concepts before proceeding to the next chapter.

The key concepts or definitions contained in this chapter are:

- Biomass concentration, X, in units of mg/L.
- Substrate (or waste) concentration, S, limiting nutrient source (usually organic carbon) in units of mg/L, usually expressed as biodegradable COD, i.e., ΔCOD.
- O_2 uptake, respiration or accumulated oxygen uptake in mg/L, a measure of the biochemical oxygen demand (BOD) of the substrate or waste.
- Specific growth rate, μ, in time^{-1}, defined according to Equation 1.3. This term is relatable to θ_c, F:M, and U (see Chapter 4).
- Specific cell decay rate, k_d, in time^{-1}.
- Cell yield, Y_t, maximum or true cell yield in mg cells/mg S, $\Delta X/\Delta S$; the cell yield unaffected by autodigestion, a constant which is a function of the waste and biomass, defined in Equation 1-2.
- Observed cell yield, Y_o, the net cell yield, i.e., the cell yield affected by autodigestion. The amount of autodigestion that occurs depends on the condition of growth. It is important under conditions of slow growth in continuous flow reactors (discussed in Chapter 2).
- Monod equation, relates μ and S for nontoxic substrates or wastes (see Equation 1.11).
- Haldane equation, relates μ and S for toxic or inhibitory wastes or substrates (see Equation 1.12).
- Maximum specific growth rate, μ_{max}, in time^{-1}, the highest specific growth rate obtainable for growth on a nontoxic substrate in the presence of excess concentrations of that substrate.
- Saturation constant, K_s, in mg/L. A shaping factor that determines the sharpness at which a plot of μ versus S approaches μ_{max} for growth on nontoxic substrates.
- Inhibition constant, K_i, in mg/L, a shaping factor that accounts for the peak and decrease in μ for increasing concentrations of S

in the Haldane equation for growth on toxic or inhibitory substrates.

- Critical substrate concentration, S^*, in mg/L. The concentration of an inhibitory substrate at which the peak in a plot of μ versus S occurs according to the Haldane equation (see Figure 1.8 and Equation 1.14).

- Critical specific growth rate, μ^*, in time^{-1}. The highest or peak specific growth rate attainable for growth on a toxic or inhibitory substrate according to the Haldane equation (see Figure 1.8 and Equation 1.13). *Note:* Do not confuse μ_{max} and μ^*; for inhibitory substrates, μ_{max} can never be observed experimentally. It must be determined analytically from growth data obtained at lower specific growth rates (see Chapter 5).

- Relating O_2 uptake (respiration) to growth and substrate removal.

  - Equation 1.20 or Equation 1.24 is used to convert O_2 uptake data to cell growth data.

  - Equation 1.21 or Equation 1.25 is used to convert O_2 uptake data to substrate removal data.

    These equations are based on the fundamental energy balance equation for aerobic metabolism, Equation 1.17. They provide the key to rapid and easy determination of numerical values of the biokinetic constants through respirometry.

- O_x, oxygen equivalent or COD of cells, mg COD/mg cells.

- R, respiration quotient, O_2 required to produce a unit of biomass, in mg O_2/mg X, determined in accordance with Equations 1.22 and/or 1.23.

REFERENCES AND SUGGESTED ADDITIONAL READING

D'Adamo, P.C., Rozich, A.F. and Gaudy, A.F. Jr. (1984). "Analysis of Growth Data with Inhibitory Carbon Sources," *Biotechnology and Bioengineering, XXVI,* pp. 397–402.

Gaudy, A.F. Jr., Rozich, A.F., Moran, N.R., Garniewski, S.T., and Ekambaram, A. (1988). "Methodology for Utilizing Respirometric Data to Assess Biodegradation Kinetics." *Proceedings, 42nd Purdue Industrial Waste Conference,* Lewis Publishers, Chelsea, Michigan, pp. 573–584.

Gaudy, A.F. Jr., Ekambaram, A., and Rozich, A.F. (1989). "A Respirometric Method for Biokinetic Characterization of Toxic Wastes," *Proceedings, 43rd Purdue Industrial Waste Conference*, Lewis Publishers, Chelsea, Michigan, pp. 35–44.

Gaudy, A.F. Jr., Ekambaram, A., Rozich, A.F. and Colvin, R.J. (1990). "Comparison of Respirometric Methods for Determination of Biokinetic Constants for Toxic and Nontoxic Wastes," *Proceedings, 44th Purdue Industrial Waste Conference,* Lewis Publishers, Chelsea, Michigan, pp. 393–403.

Gaudy, A.F. Jr., and Gaudy, E.T. (1988). *Elements of Bioenvironmental Engineering.* Engineering Press, Inc., San Jose, California, USA.

Porges, N., Jascewicz, L., and Hoover, S. (1956). "Principles of Biological Oxidation," in *Biological Treatment of Sewage and Industrial Wastes*, McCabe, B.J. and Eckenfelder, W.W., ed., Reinhold, New York, pp. 25–48.

Rozich, A.F., Gaudy, A.F. Jr., and D'Adamo, P.C. (1985). "Selection of Growth Rate Model for Activated Sludges Treating Phenol," *Water Research, 19,* pp. 481–490.

2 BASIC PRINCIPLES OF BIOREACTOR MODELING

INTRODUCTION

The kinetic principles described in Chapter 1 are universally applicable to the growth of aerobic organotrophic microorganisms regardless of the reactor in which the growth takes place. It may be a BOD bottle, a river, a lake, or a pot of soup left standing too long on the stove (or more contemporarily, in the microwave). However, this is not to say that the nature of the reactor and how it can be regulated and controlled by an outside agency (i.e., the design and operations engineers) cannot have a determining role in the rate and completion of the purification reaction represented by Equation 1.17.

In fact, it is essential that the reactor be designed to accommodate the mathematical or kinetic boundaries and assumptions as to experimental conditions under which the kinetic expressions were derived. Moreover, certain kinds of reactor systems allow one to exert control of the specific rate over and above that expressed in either the Monod or Haldane equations because one must understand that as S controls or determines specific growth rate, so too does specific growth rate determine S. Thus, engineering ways and means to control specific growth rate provide powerful tools to control concentration of S in the effluent, which is the "raison d'être" for the treatment plant. It is the reason the activated sludge process has come to such a position of supremacy when the kind of treatment process to employ is chosen.

There are obviously many different types of reactors in which to accommodate microbial growth, but in this chapter we shall discuss important concepts applicable to the type of reactor most apropos to the activated sludge process, i.e., a continuous flow, completely mixed, totally fluidized reactor. Two important concepts related to further discussion of models for activated sludge will be presented:

1. Hydraulic control of specific growth rate, μ.
2. Effect of cell recycle on specific growth rate.

The accuracy of the description of substrate removal, growth and respiration, which was shown in Figure 1.1 for batch growth systems and subsequently discussed up to this point, remains true for growth in continuous flow reactors. The difference between the batch and continuous flow systems is essentially one of hydraulics. This, we shall see, does have profound effect on the performance of the system. Let us change the reactor of Figure 1.1 by providing an influent and effluent pump (see Figure 2.1) so that wastewater can enter and exit from the reactor at the same rate, i.e., F_i (flow in) = F_e (flow exiting). Again, vigorous aeration is provided and the reactor is "completely mixed." (This latter term simply means that all of the contents, soluble and particulate, are at the same concentration in every part of the reactor.) Each component in the flow is immediately mixed in the reactor volume V and each unit of outflow is the same concentration X and S as it is in the reactor. Let us suppose that we set up a feed tank from which a waste at strength S_i is pumped into the reactor at some flow rate F and pumped out or allowed to flow out at the same rate F. Let us begin with the reactor filled with the wastewater at concentration S_i as before when we filled the reactor with

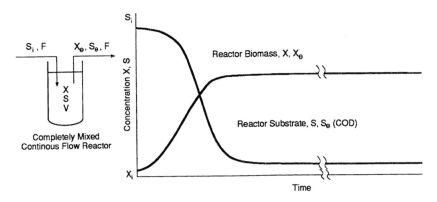

Figure 2.1. Course of substrate removal and biomass growth in a continuously fed, completely mixed reactor.

some initial concentration, S_o. We will again feed the reactor with a small concentration of acclimated activated sludge X_o, and we will again sample at various times and develop the substrate and biomass curves shown in the graph to the right of the reactor. As before, the cells grow, X increases, and the concentration of substrate in the reactor, S, decreases. Eventually, the system becomes steady with respect to concentration of S and X. That is, the concentrations remain constant regardless of how long we run the reactor. The system has come into a "steady state" with respect to S and X. The consequence of this result, which is readily observable in the reactor, can best be shown by analysis of the growth equation (Equation 1.3) and the hydraulic holding time. The hydraulic holding time is termed the reactor detention time, $\bar{t}$.

$$\bar{t} = V/F \qquad (2.1)$$

The reciprocal of $\bar{t}$ is defined as unit flow rate, D, or dilution rate.

$$D = F/V = 1/\bar{t} \qquad (2.2)$$

The term dilution rate is a convenient one, indicating that the incoming wastewater concentration S_i is diluted by the factor F/V as it enters the reactor. It is given in units of time^{-1}, i.e., the units for replacement of hydraulic liquid volume are the same as the units for specific growth rate, μ. Recall that Equation 1.3 describes the mass rate of increase for batch growth; this equation is also valid in the continuous flow reactor.

$$dX/dt = \mu X \qquad (1.3a)$$

However, in the continuous flow reactor biomass X is also exiting at a rate given by Equation 2.3.

$$dX/dt = -DX \qquad (2.3)$$

When the system attains the steady concentration in S and X, which is indicated with time in Figure 2.1, $dX/dt \to 0$, which leads to the following identity.

$$dX/dt = \mu X - DX$$
$$\mu = D \qquad (2.4)$$

Equation 2.4 is a profoundly useful one. It states that the specific growth rate is subject to hydraulic control. This is very important; since we can exert engineering control over the hydraulics of the system, i.e., $D = \mu = F/V$, we can control the specific growth rate by changing the hydraulic flow into the reactor. The equation can be refined to account for the fact that we know from experimental evidence that autodigestion can occur while the cells are growing. The specific rate of autodigestion, k_d, is usually much slower than the specific growth rate but does manifest a considerable effect on the prediction of X when the system is made to grow slowly. It should be remembered that for the activated sludge system the specific growth rate is usually held to a very low numerical value. Thus we shall refine Equation 2.4 by accounting for autodigestion and introducing the term μ_n, i.e., the net specific growth rate.

$$D = \mu_n = 1/\bar{t} = \mu - k_d \qquad (2.5)$$

Equation 2.5 represents a fundamental identity for any completely mixed continuous flow reactor system.

Although Equation 2.5 tells us much about the reactor system and its control, it is important to develop reactor equations for predicting X, S, and another important parameter, the excess cells or sludge produced in the system. Starting with the simple reactor, i.e., a reactor with no recycle of cells, such as that shown in Figure 2.1, the reactor equations are developed simply by writing the appropriate mass balances with respect to biomass X and substrate S.

$$\begin{array}{ccccc}
\text{Mass rate of} & (+)\text{ rate of change} & (-)\text{ rate of change} & (-)\text{ rate of change} \\
\text{change in X} & = & \text{due to growth} & \text{due to decay} & \text{due to outflow} \\
V\dfrac{dS}{dt} & = & V\mu X & - & Vk_d X & - & FX \quad (2.6)
\end{array}$$

$$\begin{array}{ccccc}
\text{Mass rate of} & (+)\text{ change due} & (-)\text{ change due} & (-)\text{ change due} \\
\text{change in S} & = & \text{to inflow} & \text{to outflow} & \text{to growth} \\
V\dfrac{dS}{dt} & = & FS_i & - & FS_e & - & \dfrac{\mu X}{Y_t}V \quad (2.7)
\end{array}$$

Note that for each of these mass balances (Equations 2.6 and 2.7), any biomass in the inflowing line is neglected. Also note that the substrate concentrations in the reactor and exiting the reactor are the same, i.e., $S = S_e$ in accord with the condition of complete mixing.

Dividing Equation 2.6 by V and setting $dX/dt = 0$ leads to another form of the previously given Equation 2.5.

$$\mu = D + k_d \qquad (2.5a)$$

Solving the substrate balance Equation 2.7 when $dS/dt = 0$ yields Equation 2.8.

$$X = Y_t D(S_i - S_e)/\mu \qquad (2.8)$$

ONCE-THROUGH SYSTEM (CHEMOSTAT), NONINHIBITORY MODEL

The simultaneous solution of Equations 2.5a and 2.8 is facilitated by substituting equations relating μ and S. In the case of the noninhibitory substrate, the Monod equation is substituted. These predictive equations are given in Equations 2.9 and 2.10.

$$X = \frac{Y_t D(S_i - S_e)}{D + k_d} \qquad (2.9)$$

$$S_e = \frac{K_s(D + k_d)}{\mu_{max} - (D + k_d)} \qquad (2.10)$$

The performance of the reactor with respect to biomass concentration and substrate at various hydraulically controlled growth rates (D values) is demonstrated in Figure 2.2. The effect of autodigestion on X is felt at very low D values (high $\bar{t}$ values) and as the system is run at increased values of μ, the steady state concentration of S increases until it reaches the value of S_i and the biomass is completely washed out of the reactor. The washout value of D is easily calculated. Combining Equations 1.11 and 2.5a and substituting S_i for S provides Equation 2.11.

$$D' = \mu_n' = \frac{\mu_{max} S_i}{K_s + S_i} - k_d \qquad (2.11)$$

The numerical values D' and μ_n' are termed the critical dilution rate or critical net specific growth rate when the waste substrate is of a noninhibitory nature. The term is comparable to μ^* or μ_n^* when the waste is of an inhibitory or toxic nature.

For the case shown in the figure, this washout (i.e. critical) D' or μ_n' is 0.46 hr^{-1}. This calculation tells us that biomass exhibiting the biokinetic characteristics $\mu_{max} = 0.50$ hr^{-1}, $K_s = 75$ mg/L, and $k_d = 0.005$ hr^{-1} cannot be made to grow at a net specific growth rate above 0.460 hr^{-1}. Prior to approaching this critical μ_n value, there is increasing loss of S, i.e., increasing deterioration of treatment efficiency. The flatness (or

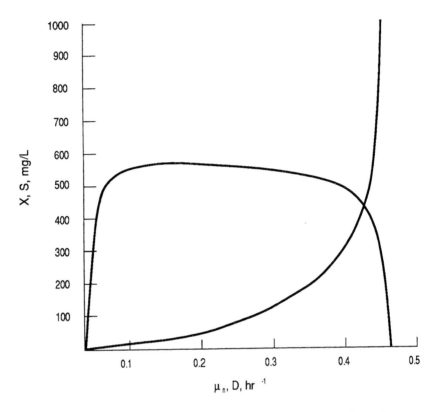

Figure 2.2. **Predicted steady-state biomass and substrate concentrations in a once-through reactor according to Equations 2.9 and 2.10. $\mu_{max} = 0.5$ hr^{-1}, $K_s = 75$ mg/L, $Y_t = 0.6$, $k_d = 0.005$ hr^{-1}, $S_i = 1,000$ mg/L (from Gaudy and Gaudy, 1988).**

sharpness) of the curve predicting effluent quality depends on the value of K_s. Lower K_s values will cause the curve to be flatter as the critical net specific growth rate, μ_n' is approached. The figure also shows that the slower one causes the system to grow, the lower is S_e, i.e., the treatment efficiency is better at lower net growth rates. Also, it is seen that for very low μ_n values, X becomes lower because of autodigestion of the biomass, i.e., μ approaches k_d.

ONCE-THROUGH SYSTEM (CHEMOSTAT), INHIBITORY MODEL

Substituting the Haldane equation for μ in Equations 2.5 and 2.8 provides the following reactor equations for inhibitory systems.

$$X = \frac{Y_t D(S_i - S_e)}{D + k_d} \qquad (2.9)$$

$$S = \left[\frac{\mu_{max}}{D + k_d} - 1 - \left[\left(1 - \frac{\mu_{max}}{D + k_d} \right)^2 - 4 \frac{K_s}{K_i} \right]^{1/2} \right] / (2/K_i) \qquad (2.12)$$

Figure 2.3 compares the reactor performance in accord with equations using the Haldane and Monod expressions, i.e., Equation 2.10 vs. Equa-

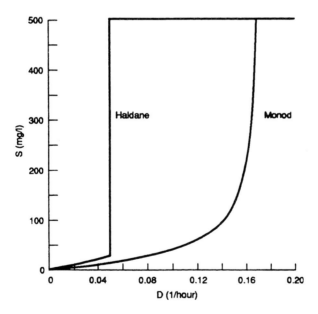

Figure 2.3. Comparison of predicted dilute-out behavior of Monod and Haldane specific growth rate relationships in a once-through (chemostat) reactor. $\mu_{max} = 0.194$ hr^{-1}, $K_s = 48$ mg/L, $K_i = 62$ mg/L, $Y_t = 1.02$ mg/mg, and $k_d = 0.02$ hr^{-1} (from Rozich and Gaudy, 1986).

tion 2.12. The behavior at increasing values of D or μ_n (decreasing $\bar{t}$) is radically different for the noninhibitory and inhibitory substrates. It is seen that the system washes out very abruptly at a much lower substrate concentration and a different dilution rate for the inhibitory substrate. There is no sliding up of S, no warning that μ_n may be approaching the critical specific growth rate, as there is with the nontoxic waste. From the biochemical constants for this system in which the cells are growing on phenol, the critical net specific growth rate μ_n^* as calculated using Equations 1.13 and 2.5 was 0.05 hr^{-1}. One can readily see that for the inhibitory waste there is greater need to control the net specific growth rate and much greater need to protect the system from sudden changes in S_i because total failure of the treatment system, i.e., washout, can occur at much lower waste substrate concentrations and it occurs more abruptly than with noninhibitory wastes.

PREDICTION OF EXCESS BIOMASS (EXCESS SLUDGE)

Thus far, we have equations that predict S and X for systems when the reactor is controlled by its operator at various values of net specific growth rate. We can only predict S and X if we know the numerical values of the biokinetic constants, μ_{max}, K_s, K_i (if the waste is inhibitory), Y_t, and k_d. These values are obtained through performance of periodic laboratory tests, easily facilitated through use of data and Equations 1.20 and 1.21 or 1.24 and 1.25. The methodology is discussed in Chapter 5.

The amount of excess sludge produced, X_w, for the once-through reactor system is readily calculated on the basis of mass per unit time from Equation 2.13.

$$X_w = VX\mu_n \tag{2.13}$$

Another important concept that can be introduced at this point is that of the time it would take to replace the biomass in the reactor. If one knew X_w, the biomass replacement time Θ_c could be readily determined by the following equation.

$$\Theta_c = VX/X_w \tag{2.14}$$

This term is familiar to many engineers as the sludge age or mean cell residence time (MCRT). Examination of Equations 2.13 and 2.14 show that μ_n and Θ_c are related as shown in Equation 2.15.

$$\mu_n = 1/\Theta_c \qquad (2.15)$$

This relationship is discussed more fully in Chapter 3.

REVIEW

In Chapter 1 we introduced the concept of mass and energy balances based on the partition of substrate into that which is used for respiration (O_2 uptake) and that channelled into new biomass. This permits one to develop a useful quantitative relationship for determining X and/or S using O_2 uptake, which is readily obtained in laboratory tests on a waste/biomass (sludge) system. In Chapter 2, the growth relationships (Monod, Haldane) have been employed in simple mass balance equations for a bioreactor (once-through [chemostat] flow system) and predictive equations for S, X, and X_w were developed; the relationship between the net specific growth rate μ_n and mean cell residence time Θ_c has been shown. It has been demonstrated that the μ_n (or Θ_c) profoundly affects the performance of the system because this parameter controls the concentration of substrate exiting the reactor.

There is still one last important engineering expedient that must be delineated. The recycle of biomass to the reactor provides a powerful tool for controlling reactor behavior because it offers two additional ways to control the net specific growth rate of the system.

EFFECT OF BIOMASS RECYCLE

A simple once-through reactor system is converted into one capable of recycling biomass by adding a cell separator after the reactor and making provisions to recycle concentrated cells. In the case of the activated sludge process, the cell separator is a sedimentation basin. The essential difference between the systems is delineated in the schematic diagram of Figure 2.4.

The reactor equations in X and S for the once-through system are different from the ones for the recycle reactor simply because the reactor now has two input lines, the wastewater and the recycle line. This addi-

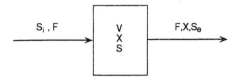

Once Through Reactor System

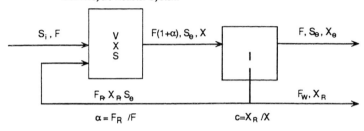

Cell Recycle Reactor System

Figure 2.4.   Comparison of flow diagram for once-through and recycle reactor systems.

tional input consists of a recycle flow F_R which can be controlled at some fraction (α) of the inflowing wastewater ($\alpha = F_R/F$). This recycle flow contains a biomass concentration (X_R), which may be controlled at some ratio c (c $= X_R/X$). The flow may contain some amount of substrate, and for purposes of simplification we may assume that there is essentially no chemical or biochemical activity in the cell separator or clarifier so that the substrate concentration in both the effluent and the recycle flow are the same as in the reactor.

Mass balances for dX/dt and dS/dt are given by Equations 2.16 and 2.17.

$$V \frac{dX}{dt} = (\alpha F)(cX) + V\mu X - Vk_d X - (1 + \alpha)FX \qquad (2.16)$$

$$\phantom{V \frac{dX}{dt} = } \underset{\text{inflow}}{} \quad \underset{\text{growth}}{} \quad \underset{\text{decay}}{} \quad \underset{\text{outflow}}{}$$
(recycle)

$$V \frac{dS}{dt} = FS_i + \alpha FS_e - \frac{\mu XV}{Y_t} - F(1 + \alpha)S_e \qquad (2.17)$$

$$\phantom{V \frac{dS}{dt} = } \underset{\text{inflow}}{} \quad \underset{\text{recycle}}{} \quad \underset{\text{utilization}}{} \quad \underset{\text{outflow}}{}$$

In the example below it is assumed that the waste is of a nontoxic nature and the Monod relationship between μ and S has been employed. As before, we divide through by V and let dX/dt and dS/dt approach zero. Solution of Equation 2.16 yields Equation 2.18.

$$\mu = D(1 + \alpha - \alpha c) + k_d \qquad (2.18)$$

We see that, as in Equation 2.5, μ still falls under hydraulic control; that is, engineering control, but (D or $\bar{t}$) is not the only hydraulic control. The recycle flow ratio α and the recycle sludge concentration ratio c also exert controls over the specific growth rate. As for the once-through system, solving the substrate balance for the steady-state condition and substituting the Monod relationship for μ leads to the following predictive equations for X and S_e.

$$X = \frac{Y_t D(S_i - S_e)}{D(1 + \alpha - \alpha c) + k_d} \qquad (2.19)$$

$$S_e = \frac{K_s[D(1 + \alpha - \alpha c) + k_d]}{\mu_{max} - [D(1 + \alpha - \alpha c) + k_d]} \qquad (2.20)$$

It is seen that Equations 2.9 and 2.10 for the once-through (chemostat) system differ from the above equations only by the recycle factor $(1 + \alpha - \alpha \cdot c)$. We know from previous discussions that S_e, effluent quality, is determined by μ and from Equation 2.18 that μ is determined largely by the factor $D(1 + \alpha - \alpha c)$. We can see the tremendous effect recycle has on the effluent quality by examination of Figure 2.5.

For this demonstration the sludge concentration factor (c) of 4 was selected since this is a concentration ratio easily obtained by simple sedimentation of cells and is commonly found in activated sludge systems. The most striking conclusion one draws from this figure is that cell recycle enables one to produce much lower S_e values for less reactor detention time, i.e., less reactor volume than is possible without recycling. Recycle also provides the opportunity to control S_e after the plant is in operation by allowing the operator to select and control the numerical values of α and c. The only way D (i.e., $\bar{t}$) can change is by a change in F, since V is fixed once the plant is built. Cyclical as well as long-term increases in F only serve to worsen effluent quality because an increase in F obviously increases μ. Thus, α and c provide important engineering controls. These along with the biokinetic constants μ_{max}, K_s, K_i, Y_t, and k_d exert an effect on μ and thus on the effluent quality produced. As we

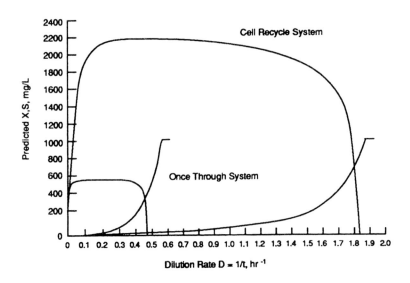

Figure 2.5. Comparisons of predicted dilute-out behavior for once-through (chemostat) and cell recycle reactors. μ_{max} = 0.5 hr⁻¹, K_s = 75 mg/L, Y_t = 0.6, K_d = 0.005 hr⁻¹, α = 0.25, c = X_R/X = 4 (from Gaudy and Gaudy, 1988).

shall learn in a succeeding chapter, the value of S_i can also exert a controlling effect on μ and consequently on the quality of the effluent. Controlling of effluent quality is, of course, the aim of the exercise of design and operation of the activated sludge process; the engineering and biokinetic factors are the tools available to accomplish the job of wastewater purification. Other important aspects, such as provision of adequate mixing and oxygen and separation of biomass from the effluent in the clarifier, will not be addressed in detail in subsequent chapters of this text.

Thus far we have discussed basic kinetic considerations that led to the model Equations 2.9 and 2.10. One may question the utility of these reactor equations. Do they have predictive value and can they be used to design and operate an activated sludge process? These equations, without consideration of k_d, were developed many years ago to describe the growth of various species of microorganisms. Almost as many years ago, they were employed in experimental laboratory activated sludge processes to determine their suitability for mixed cultures and substrate systems. It was found that once-through reactor (chemostat) behavior was predicted rather well by Equations 2.8 and 2.9 using values of the biokinetic constants determined from independent batch tests of the system. For the recycle system, Equations 2.19 and 2.20 were found to be acceptable, but they were not readily applicable in view of accepted engineering practice in the pollution control field. One could easily employ α, F_R/F, as a control because it was easily measured and pumps are rather easily controlled. Thus, a desired α could be selected and maintained at that value. However, the concentration factor c is not so selectable, nor does X_R remain constant. Most clarifier underflow concentrations vary somewhat because of changes in sludge settling characteristics. Also, X_R is a variable depending on F and F_R. In practice, it was found that maintaining c at some selected value was an extremely difficult task. It was also found that the steady-state condition of X was subject to some variations because slight changes in Y_t, K_s, K_i, μ_{max}, and k_d were exaggerated due to the requirement to hold c constant. All these factors led to the conclusion that c was not a very good engineering control variable. In order to accommodate the equation to engineering reality, X_R, rather than c, was used along with α and D, i.e., Equation 2.18 was rewritten:

$$\mu = D\left(1 + \alpha - \alpha \frac{X_R}{X}\right) + k_d \qquad (2.21)$$

This change does not fundamentally modify the theory of continuous culture as presented in this chapter, but it does have a fundamental effect on the form of the final engineering equations. One cannot simply substitute an X_R/X term for c in Equations 2.19 and 2.20. It is necessary to make this substitution in Equation 2.16 and then to proceed with the simultaneous solution of Equations 2.16 and 2.17. The derivation is presented in Chapter 3, leading to the development of engineering equations. There it will be shown that in addition to α, X_R, D (or $\bar{t}$), k_d, Y_t, μ_{max}, K_s, and K_i, the influent substrate concentration S_i also influences μ and thus S and X. The final engineering equations relate all of the important factors governing the quality of the effluent. They have been extensively tested and, as will be shown in case studies presented in Chapter 7, found to be of excellent predictive value for large-scale wastewater treatment facilities.

KEY CONCEPT SUMMARY

The key concepts which are contained in this chapter are:

- In addition to the fundamental definitions and relationships between growth, substrate removal rate, and respiration (O_2 uptake) that govern all aerobic microbial growth, presented in Chapter 1, it is also important to note that the reactor and/or reactor configuration has a significant impact on growth kinetics.
- The important reactor concepts are the hydraulic control of μ (Equation 2.4) and the additional hydraulic control of μ by biomass recycle (Equation 2.18).
- The recycle sludge concentration, X_R, rather than c (X_R/X), is the preferred control parameter for field systems. Equation 2.21 is vital to the development of engineering equations, which are derived and discussed in Chapter 3.

REFERENCES AND SUGGESTED ADDITIONAL READING

Gaudy, A.F. Jr., Ramanathan, M. and Rao, B.S. (1967). "Kinetic Behavior of Heterogeneous Populations in Completely Mixed Reactors," *Biotech. Bioeng., 9*, pp. 387–411.

Gaudy, A.F. Jr., and Srinivasaragahavan, R. (1974). "Experimental Studies on a Kinetic Model for Design and Operation of Activated Sludge Processes," *Biotech. Bioeng., 16*, pp. 723–728.

Gaudy, A.F. Jr., and Gaudy, E.T. (1988). *Elements of Bioenvironmental Engineering,* Engineering Press, Inc., San Jose, California.

Ramanathan, M., and Gaudy, A.F. Jr. (1969). "Effect of High Substrate Concentration and Cell Feedback on Kinetic Behavior of Heterogeneous Populations in Completely Mixed Systems," *Biotech. Bioeng., 11*, pp. 207–237.

Rozich, A.F., and Gaudy, A.F. Jr. (1986). "Process Technology for the Biological Treatment of Toxic Organic Wastes," *Hazardous and Industrial Waste Testing and Disposal, 6,* ASTM STP 933, pp. 319–333.

Srinivasaragahavan, R., and Gaudy, A.F. Jr. (1975). "Operational Performance of an Activated Sludge Process with Constant Sludge Feedback," *J. Water Poll. Control Fed., 47*, pp. 1946–1960.

3 ENGINEERING MODELS FOR ACTIVATED SLUDGE SYSTEMS

INTRODUCTION

Predictive models for activated sludge systems are derived using the same methodology detailed in Chapter 2 regarding the theory of continuous culture. This is essentially a reactor engineering approach. The basic procedure is to write mass balance equations for biomass and substrate around the reactor. An appropriate expression for relating growth rate to substrate concentration is then inserted into the mass balance equations, which then are solved simultaneously to obtain predictors for X and S, the reactor biomass and substrate concentrations, respectively. This same procedure can be used to obtain equations for any configuration of an activated sludge system. In this chapter, we will present the derivation of predictive equations for an activated sludge system with one completely mixed reactor. We will also present the derivation of predictive equations for an activated sludge system with multiple reactors.

It is important to emphasize that the treatment performance of an activated sludge system is largely controlled via specific growth rate. Growth rate (and its related environmental counterparts, μc, F:M, etc.) is in turn determined by engineering controls such as flow rate or detention time, recycle sludge concentration, and recycle flow rate. The influent waste concentration also exerts an influence on reactor growth rate. The activated sludge models are governed by two main sets of constants or controls: biokinetic and engineering. The biokinetic growth constants (those found in Equations 1.11 and 1.12) quantify the capability of a reactor's biomass to degrade a target waste. The engineering constants are physical components that the designer or operator can control.

39

(Influent waste strength can be controlled with techniques such as equalization.) The model's purpose is to link the biokinetic constants, which are largely a function of environmental conditions, with the engineering constants, which are under the control of the environmental technologist. If the waste is difficult to degrade or inhibitory (as indicated by the values of the biokinetic constants), the model will quantify those values of the engineering parameters which are needed to maintain reliable process performance for the target treatment condition.

This chapter contains three main sections. The first section deals with the derivation of the predictive equations for activated sludge systems. Models are presented for both noninhibitory (Monod function) and inhibitory (Haldane function) wastes. The second section describes the derivation of predictive equations for multiple-reactor (tanks-in-series) systems. The purpose of this section is to demonstrate that the modeling approach presented herein is flexible and that deriving a model for a specific system or reactor configuration is simply a matter of following the methodology given in this chapter. The last section concerns critical point analysis for treatment of inhibitory wastes. Many toxic and hazardous wastes are characterized by inhibition kinetics. As noted in Chapter 2, wastes that cause an inhibition response from the biomass require special consideration when designing and operating a bioreactor. This is primarily due to the potential to exceed initial operating conditions, which leads to process failures for biological systems treating inhibitory wastes. Quantitative methods to calculate the operational location of critical operating points are presented in the third section of this chapter.

DERIVATION OF PREDICTIVE EQUATIONS

Noninhibitory Wastes

A flow sheet of an activated sludge system with one completely mixed reactor is given in Figure 3.1. It is assumed that the reactor maintains aerobic conditions and that recycle sludge concentration, X_R has a significant impact on reactor performance. X_R can be held constant or can be controlled at a value selected by the engineer. Although this is rarely performed in field practice, it needs to be emphasized that this parameter imparts a substantial impact on reactor growth rate. A highly fluctuating recycle sludge concentration will lead to process upsets in difficult treat-

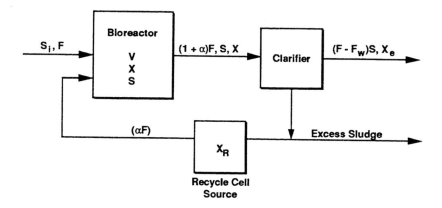

Figure 3.1.   Flow diagram used for developing model equations.

ment situations unless other compensatory measures (e.g., long reactor detention times) are implemented.

The appropriate mass balance equations for biomass, X, and soluble substrate, S, are given below:

$$V \frac{dX}{dt} = \alpha F X_R + \mu X V - k_d X V - (1 + \alpha) F X \qquad (3.1)$$

$$V \frac{dS}{dt} = F S_i + \alpha F S_R - (1 + \alpha) F S_e - \frac{\mu X V}{Y_t} \qquad (3.2)$$

Since in the steady state, dX/dt and dS/dt = 0, an algebraic solution is possible, and using t = V/F, Equation 3.1 can be solved for μ.

$$\mu = D \left(1 + \alpha - \alpha \frac{X_R}{X}\right) + k_d \qquad (3.3)$$

Also note that since

$$\mu = \mu_n + k_d \qquad (3.4)$$

$$\mu_n = D \left(1 + \alpha - \alpha \frac{X_R}{X}\right) \qquad (3.5)$$

Assuming steady-state conditions, Equation 3.2 can be solved for X:

$$X = \frac{Y_t}{\mu} [DS_i + \alpha DS_R - (1 + \alpha)DS_e].$$
(3.6)

To obtain a predictive equation for S, Equation 3.6 is substituted into Equation 3.3:

$$\mu = [(1 + \alpha)D + k_d] - \frac{\mu \alpha X_R}{Y (S_i + \alpha S_R - (1 + \alpha)S_e)}$$
(3.7)

Factoring out μ,

$$\mu \left[1 + \frac{\alpha X_R Y_t}{S_i + \alpha S_R - (1 + \alpha)S_e} \right] = [(1 + \alpha)D + k_d]$$
(3.8)

To obtain a predictive equation for noninhibitory substrates, substitute the Monod equation (Equation 1.11) in place of μ:

$$\frac{\mu_{max} S_e}{S_e + K_s + S_e^2/K_i}[S_i + \alpha S_R - (1 + \alpha)S_e]$$
$$+ \left(\frac{\mu_{max} S_e}{S_e + K_s + S_e^2/K_i} \right) \frac{\alpha X_R}{Y_t}$$
$$= [(1 + \alpha)D + k_d][S_i + \alpha S_R]$$
$$- [(1 + \alpha)D + k_d](1 + \alpha)S_e$$
(3.9)

After collecting terms, the predictive equation for S is:

$$S_e = \frac{-b \pm (b^2 - 4ac)^{1/2}}{2a}$$
(3.10)

$$a = ([(1 + \alpha)D + k_d] - \mu_{max})(1 + \alpha)$$
$$b = \mu_{max} \left(S_i + S_R + \frac{\alpha X_R}{Y_t} \right)$$
$$\quad + [(1 + \alpha)D + k_d][(1 + \alpha)K_s - (S_i + \alpha S_R)]$$
$$c = - (S_i + \alpha S_R)[(1 + \alpha)D + k_d]K_s$$

A predictive equation for X, the reactor biomass concentration, is generated by combining Equations 3.3 and 3.6:

$$X = \frac{Y_t (S_i + \alpha S_R - (1 + \alpha)S)D}{\left[\frac{1}{t} (1 + \alpha - \alpha\frac{X_R}{X}) + k_d\right]} \tag{3.11}$$

$$X = \frac{Y_t (S_i + \alpha S_R - (1 + \alpha)S)}{1 + \alpha - \alpha\frac{X_R}{X} + k_d/D} \tag{3.12}$$

$$(1 + \alpha + k_d/D)X = Y_t (S_i + \alpha S_R - (1 + \alpha)S) + \alpha X_R \tag{3.13}$$

$$X = \frac{Y_t (S_i + \alpha S_R - (1 + \alpha)S) + \alpha X_R}{1 + \alpha + k_d/D} \tag{3.14}$$

An equation for waste sludge production, X_w, is derived by recognizing that the amount of sludge that is wasted (or produced) is the difference between the amount that exits the reactor and the amount that enters the reactor:

$$X_w = (1 + \alpha)FX - \alpha FX_R \tag{3.15}$$

Equation 3.15 can be rewritten as follows:

$$X_w = \frac{FX}{V} (1 + \alpha - \alpha \frac{X_R}{X}) V \tag{3.16}$$

$$X_w = D \left(1 + \alpha - \alpha \frac{X_R}{X}\right) XV \tag{3.17}$$

$$X_w = \mu_n XV \tag{3.18}$$

Equation 3.18 will predict the amount of excess sludge that will be produced by the activated sludge system.

Inhibitory Wastes

Predictive equations for inhibitory substrates such as those that are often characteristic of toxic or hazardous wastes are derived by substituting the Haldane equation, in lieu of the Monod equation, into Equation 3.8. After collecting terms, the predictive equation for S, or effluent quality, is obtained:

$$aS_e^3 + bS_e^2 + cS_e + d = 0$$

$$a = \frac{[(1 + \alpha) D + k_d] (1 + \alpha)}{K_i}$$

$$b = [(1 + \alpha)D + k_d]\left[(1 + \alpha)D - \frac{S_i + \alpha S_R}{K_i}\right]$$

$$- \mu_{max}(1 + \alpha)$$

$$c = [(1 + \alpha)D + k_d][(1 + \alpha)K_s - (S_i + \alpha S_R)]$$

$$+ \mu_{max}\left(S_i + \alpha S_R + \frac{\alpha X_R}{Y_t}\right)$$

$$d = -[(1 + \alpha)D + k_d][S_i + \alpha S_R]K_s$$

(3.19)

Although Equation 3.19 is a cubic, it is easily solved with a numerical method such as a Newton-Raphson iteration technique. The predictive equations for X and X_w, the reactor biomass concentration and waste sludge production rate, respectively, are the same for both noninhibitory and inhibitory wastes.

Equations for both the noninhibitory and inhibitory wastes are summarized in Table 3.1. It is evident in this table that the only difference between the noninhibitory and inhibitory models is the inhibition constant, K_i. It is interesting to note that, as the inhibition constant tends to go to infinity, which is indicative of a more noninhibitory waste, the inhibitory equations reduce to the noninhibitory form. Thus, one can view the inhibitory predictive equations as addressing a more general case for waste treatment, since they incorporate the impact of waste, or substrate, inhibition on process performance.

As an example of the application of the predictive equations, consider Figure 3.2. This figure shows their predictive curves for effluent quality S and reactor biomass concentration X for an activated sludge reactor treating phenol. In this case, both the noninhibitory and inhibitory models were utilized to make predictions of effluent quality. The output from the modeling procedure is a prediction of effluent quality. The input to the model consists of the values of the biokinetic constants and the values of the engineering control parameters, reactor detention time (or influent flow rate), recycle sludge concentration, and recycle flow ratio. Influent waste strength, or concentration, is also an input parameter. The curves in Figure 3.2 are generated using the equations from Table 3.1 and provide predictions of effluent quality in terms of soluble biodegradable COD for the specified values of the biokinetic and engineering constants and the influent waste strength.

Table 3.1 Steady-State Predictive Equations for Activated Sludge Systems

Inhibitory Substrate (Haldane)	Noninhibitory Substrate (Monod)
S_e: $\quad aS_e^3 + bS_e^2 + cS_e + d = 0 \qquad (3.19)$	$S_e = \dfrac{-b \pm (b^2 - 4ac)^{1/2}}{2a} \qquad (3.10)$
$a = \dfrac{(1+\alpha)[(1+\alpha)D + k_d]}{K_i}$	$a = \left([(1+\alpha)D + k_d] - \mu_{max}\right)(1+\alpha)$
$b = [(1+\alpha)D + k_d]\left(1 + \alpha - \dfrac{(S_i + \alpha S_R)}{K_i}\right) - (1+\alpha)\mu_{max}$	$b = \mu_{max}\left(S_i + S_R + \dfrac{\alpha X_R}{Y_t}\right) + [(1+\alpha)D + k_d][(1+\alpha)K_s - (S_i + \alpha S_R)]$
$c = \mu_{max}\left((S_i + \alpha S_R) + \dfrac{\alpha X_R}{Y_t}\right) + [(1+\alpha)D + k_d][(1+\alpha)K_s - (S_i + \alpha S_R)]$	$c = -(S_i + \alpha S_R)[(1+\alpha)D + k_d]K_s$
$d = -(S_i + \alpha S_R)[(1+\alpha)D + k_d]K_s$	
$X = \dfrac{Y_t[(S_i + \alpha S_R) - (1+\alpha)S_e] + \alpha X_R}{1 + \alpha + k_d/D} \qquad (3.12)$	$X = \dfrac{Y_t[S_i + \alpha S_R - (1+\alpha)S_e] + \alpha X_R}{1 + \alpha + k_d/D} \qquad (3.12)$
$X_w = VX\mu_n \qquad (3.18)$	$X_w = VX\mu_n \qquad (3.18)$

Note: S_e = S in completely mixed systems. $D = 1/t$.

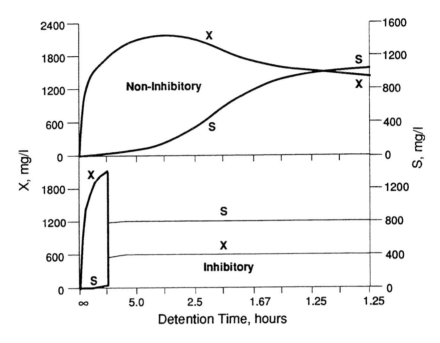

Figure 3.2. Comparison of predictive curves for effluent quality calculated from the Monod and Haldane equations for a constant X_R activated sludge process treating phenol at an S_i of 1500 mg/L. Values of parameters and biokinetic constants are: $X_R = 6,000$ mg/L, $\alpha = 0.25$, $\mu_{max} = 0.19$ hr^{-1}, $K_s = 48$ mg/L, $K_i = 62$ mg/L, $Y_t = 1.02$, and $k_d = 0.47$ days^{-1}.

Figure 3.2 also makes an important point regarding the use of the inhibitory model. It is clear that the inhibitory model for phenol predicts a sudden decrease in effluent quality once a certain operating condition is attained. This condition corresponds to the peak of the Haldane equation as discussed in Chapter 2 and is not predicted with the noninhibitory model. Extensive testing of the model has shown that the behavior of activated sludge systems treating inhibitory wastes, i.e., those wastes that are characterized with Haldane kinetics, is accurately predicted by this model. It is clear in Figure 3.2 that the Monod-based model predicts a gradual loss in performance capacity. Consequently, failure to account for the impact of the inhibition kinetics on the design and operation of the biological treatment system leads to a gross overestimate of the effective operating range. The bottom line is that activated sludge systems that are dealing with inhibitory wastes must be modeled using an algorithm that accounts for the impact of inhibition. The alternative is to design or operate the system using an extremely conservative approach (oversized aeration basins) or to risk system failure by exceeding the system's critical operating point.

APPLICATION FOR MULTIPLE REACTOR SYSTEMS

Many activated sludge systems are single, completely mixed basins. However, other systems also use multiple reactors or other different configurations. The objective of this section is to demonstrate the flexibility of the modeling approach presented herein for addressing different reactor configurations. That is, if a system is different than complete mix, one can derive the equations that apply for that system. As an example, we present the derivation of equations for a multiple-cell activated sludge system consisting of "m" number of cells as depicted in Figure 3.3. The approach is the same as that used for the single cell system except that multiple reactors now must be modeled.

The multi-reactor model is derived by writing mass balance equations for biomass and substrate concentrations for each cell. For the first cell, the mass balance equations are:

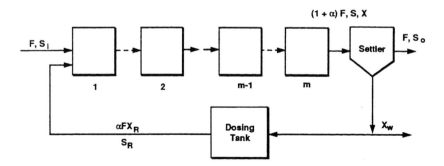

Figure 3.3.   Flow schematic of multi-cell activated sludge system.

$$\frac{V}{m} \frac{dX(1)}{dt} = \alpha FX_R + \mu(1)X(1)\frac{V}{m} - k_d X(1)\frac{V}{m} - (1 + \alpha)FX(1) \quad (3.20)$$

$$\frac{V}{m} \frac{dS(1)}{dt} = FS_i + \alpha FS_R - (1 + \alpha)FS(1) - \frac{\mu(1)}{Y_t} X(1)\frac{V}{m} \quad (3.21)$$

At steady state, the time derivatives equal zero, and Equations 3.20 and 3.21 become:

$$0 = \alpha mDX_R = \mu(1)X(1) - k_d X(1) - (1 + \alpha)mDX(1) \quad (3.22)$$

$$0 = mDS_i + \alpha mDS_R - (1 + \alpha)mDS(1) - \frac{\mu(1)X(1)}{Y_t} \quad (3.23)$$

In Equations 3.22 and 3.23, m represents the number of completely mixed cells in the aeration tank and V is the total aeration tank volume. By utilizing equally sized cells, the volume of each cell is V/m. $X_{(1)}$ and $S_{(1)}$ are the biomass and substrate concentrations in reactor 1, respectively; α is the recycle flow ratio; D is the total tank dilution rate time, F/V; X_R is the recycle sludge concentration; $\mu_{(1)}$ is the specific growth rate in reactor 1; k_d is the specific decay rate; S_i is the influent substrate concentration to the tank; S_R is the substrate concentration in the recycle line (usually assumed to equal zero); and Y_t is the true cell yield.

Steady-state solutions are obtained by utilizing an appropriate function for relating μ to S and then solving Equations 3.22 and 3.23 for $X_{(1)}$ and $S_{(1)}$. As done previously, the Monod equation is used for noninhibitory wastes and the Haldane equation is employed for toxic or inhibitory wastes. The steady-state predictive equations for reactor 1 for both noninhibitory and inhibitory wastes are given in Table 3.2.

Mass balance equations for the other cells in the multiple-cell aeration tank, i.e., reactors 2 through m, are given in Equations 3.24 and 3.25:

$$\frac{V}{m} \frac{dX(j)}{dt} = (1 + \alpha)FX(j-1) + \mu(j)X(j)\frac{V}{m} - k_d X(j)\frac{V}{m} - (1 + \alpha)FX(j) \quad (3.24)$$

$$\frac{V}{m} \frac{dS(j)}{dt} = (1 + \alpha)FS(j-1) - (1 + \alpha)FS(j) - \frac{\mu(j) X(j)}{Y_t} \frac{V}{m} \quad (3.25)$$

It should be noted that Equations 3.24 and 3.25 apply for cells 2 through m provided that $X_{(j-1)}$ and $S_{(j-1)}$ are already calculated. The steady-state versions of these equations are given in Equations 3.26 and 3.27:

Table 3.2 Steady-State Predictive Equations for Activated Sludge Cells-In-Series System

Noninhibitory Substrates (Monod)	Inhibitory Substrates (Haldane)
Reactor 1	Reactor 1

Noninhibitory Substrates (Monod) — Reactor 1

$$S(1) = \frac{-b \pm (b^2 - 4ac)^{1/2}}{2a} \qquad (3.27a)$$

$$a = (((1 + \alpha)mD) - \mu_{max})(1 + \alpha)$$

$$b = \mu_{max}\left(S_i + S_R + \frac{\alpha X_R}{Y_t}\right) + ((1 + \alpha)mD + k_d)((1 + \alpha)K_s - (S_i + \alpha S_R)$$

$$c = -(S_i + \alpha S_R)((1 + \alpha)mD + k_d)K_i$$

$$X(1) = \frac{Y_t[S_i + \alpha S_R - (1 + \alpha)S(1)] + \alpha X_R}{1 + \alpha + \dfrac{k_d}{mD}} \qquad (3.26a)$$

Inhibitory Substrates (Haldane) — Reactor 1

$$a\,S^3(1) + b\,S^2(1) + c\,S(1) + d = 0 \qquad (3.27c)$$

$$a = \frac{((1 + \alpha)mD + k_d)}{K_i}$$

$$b = ((1 + \alpha)mD + k_d)\left(1 - \left(\frac{S_i + \alpha S_R}{K_i}\right)\right) - (1 + \alpha)\mu_{max}$$

$$c = \mu_{max}\left((S_i + \alpha S_R) + \frac{\alpha X_R}{Y_t}\right) + [(1 + \alpha)mD + k_d]((1 + \alpha)K_s - (S_i + \alpha S_R))$$

$$d = -(S_i + \alpha S_R)((1 + \alpha)mD + k_d)K_s$$

$$X(1) = \frac{Y_t[S_i + \alpha S_R - (1 + \alpha)S(1)] + \alpha X_R}{1 + \alpha + \dfrac{k_d}{mD}} \qquad (3.26a)$$

Table 3.2 continued

Noninhibitory Substrates (Monod)	Inhibitory Substrates (Haldane)

Noninhibitory Substrates (Monod)

Reactors 2 through m

$$S(j) = \frac{-b \pm (b^2 - 4ac)^{1/2}}{2a} \qquad (3.27b)$$

$$a = ((1 + \alpha)mD + k_d) - \mu_{max}$$

$$b = \mu_{max}\left(S(j-1) + \frac{X(j-1)}{Y_t}\right) + ((1 + \alpha)mD + k_d)(K_s - S(j-1))$$

$$c = -((1 + \alpha)mD + k_d)K_sS(j-1)$$

$$X(j) = \frac{mD(1 + \alpha)X(j-1)}{mD(1 + \alpha) + k_d - \left(\dfrac{\mu_{max}S(j)}{K_s + S(j)}\right)} \qquad (3.26b)$$

$$X_w = F((1 + \alpha)X(m) - \alpha X_R) \qquad (3.26c)$$

Inhibitory Substrates (Haldane)

Reactors 2 through m

$$a\,S^3(j) + b\,S^2(j) + c\,S(j) + d = 0 \qquad (3.27d)$$

$$a = \frac{((1 + \alpha)mD + k_d)}{K_i}$$

$$b = ((1 + \alpha)mD + k_d)\left(1 - \frac{S(j-1)}{K_i}\right) - \mu_{max}$$

$$c = ((1 + \alpha)mD + k_d)(K_s - S(j-1)) + \mu_{max}\left(S(j-1) + \frac{X(j-1)}{Y_t}\right) +$$

$$d = -((1 + \alpha)mD + k_d)K_sS(j-1)$$

$$X(j) = \frac{mD(1 + \alpha)X(j-1)}{mD(1 + \alpha) + k_d - \left(\dfrac{\mu_{max}S(j)}{K_s + S(j) + S^2(j)/K_i}\right)} \qquad (3.26b)$$

$$X_w = F((1 + \alpha)X(m) - \alpha X_R) \qquad (3.26c)$$

Note: $t = 1/D$.

$$0 = (1 + \alpha)mDX(j-1) + \mu(j)X(j) - k_dX(j) - (1 + \alpha)mDX(j) \quad (3.26)$$

$$0 = (1 + \alpha)mDS(j-1) - (1 + \alpha)mDS(j) - \frac{\mu(j)X(j)}{Y_t} \quad (3.27)$$

A solution technique analogous to that presented for the single, completely mixed reactor system is employed to obtain steady-state predictive equations for $X_{(j)}$ and $S_{(j)}$. These predictive equations are also given for noninhibitory and inhibitory wastes in Table 3.2. The equations in this table are applicable to an activated sludge system with m number of equal volume reactors. They also cover the case of a single, completely mixed reactor; i.e., when m = 1, the equations are the same as those presented in Table 3.1. Finally, a predictive equation is presented for waste sludge production in the multiple-cell system; this is derived by performing a mass balance for biomass production in the aeration tank.

CRITICAL POINT ANALYSIS FOR TREATMENT OF INHIBITORY WASTES

The significance of the peak of the Haldane curve or μ^* in biological reactors treating toxic or inhibitory wastes is simply that, once the reactor attains this growth rate, it is subject to sudden effluent deterioration and washout. This feature of inhibitory waste treatment was demonstrated analytically in Figure 3.2 using the model equations for an activated sludge system (single, completely mixed reactor) treating phenol. It was also noted in Chapter 2 that chemostats treating toxics wash out once the detention time produces a growth rate corresponding to μ^* in the chemostat. These features of inhibitory waste treatment make it imperative to operate reactors at growth rates that do not approach μ^*.

It needs to be stressed that the phenomenon of reactor failure at μ^* has been demonstrated often in bench-scale pilot plant systems. Figure 3.4 provides a dramatic illustration of the consequences of exceeding μ^* in an activated sludge reactor treating the toxic organic phenol. In this activated sludge system, the influent waste strength S_i was 2000 mg/L phenol, and it can be seen that, once μ^* was exceeded, effluent quality rapidly deteriorated and recovery was not possible.

It should be noted that for this activated sludge pilot plant, conditions for reactor failure were predicted using biokinetic constants determined from batch tests that used the reactor biomass and the influent phenol waste. Consequently, it is feasible to formulate operating policies that

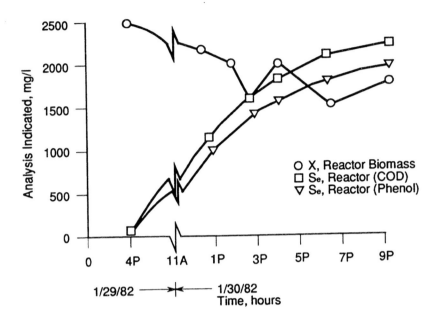

Figure 3.4.   Consequences of exceeding μ^* in activated sludge reactor treating phenol.

permit maximum treatment efficiency while avoiding reactor failure because of operation near μ^*.

Another way to formulate operating strategies for activated sludge system treating toxics is to quantify the operating conditions, or critical operating points, that result in a growth rate of μ^* in an activated sludge reactor. Equations that quantify the critical operating point are derived by considering the activated sludge process flow diagram shown in Figure 3.1. Mass balances for X and S, the reactor biomass and substrate concentrations, respectively, are written around the single, completely mixed reactor.

$$V \frac{dX}{dt} = \alpha F X_R + \mu XV - k_d XV - (1 + \alpha)FX \qquad (3.1)$$

$$V \frac{dS}{dt} = FS_i + \alpha FS_R - (1 + \alpha)FS_e - \frac{\mu XV}{Y_t} \qquad (3.2)$$

By assuming steady-state conditions and by letting D = F/V, the nominal dilution rate, Equations 3.28 and 3.29 are obtained.

$$0 = \alpha DX_R + \mu X - k_d X - (1 + \alpha)DX \qquad (3.28)$$

$$0 = DS_i + \alpha DS_R - (1 + \alpha)DS - \frac{\mu X}{Y_t} \qquad (3.29)$$

Equations 3.28 and 3.29 are manipulated to yield expressions for μ and X.

$$\mu = D\left(1 + \alpha - \frac{\alpha X_R}{X}\right) + k_d \qquad (3.30)$$

$$X = \frac{Y_t}{\mu}[DS_i + \alpha DS_R - (1 + \alpha)DS_e]. \qquad (3.6)$$

An expression for the critical detention time, t*, which produces a growth rate equal to μ* in the activated sludge reactor, is obtained by combining Equations 3.6, 1.13, 1.14, and 3.30. This expression for t* is given in Equation 3.31.

$$t^* = (1 + \alpha)\left[\left[\frac{\mu_{max}}{1 + 2\sqrt{\dfrac{K_s}{K_i}}}\right]\left[1 + \frac{\dfrac{\alpha X_R}{Y_t}}{S_i + \alpha S_R - (1 + \alpha)\sqrt{K_s K_i}}\right] - k_d\right]^{-1} \qquad (3.31)$$

It should be noted that Equation 3.31 gives a value for the detention time at which the reactor attains the critical growth rate, μ*, i.e., the detention time at which one can expect rapid effluent deterioration for an activated sludge system treating an inhibitory waste. Three engineering control variables, $\bar{t}$ (or flow rate), α, and X_R, can be selected by the designer and operator; S_i, the influent waste strength, can often be held relatively constant using equalization techniques. The bottom line is that by determining the value of the critical growth rate, μ*, via biokinetic testing using respirometric methods, one can use the model to quantify the values of the engineering controls needed to avoid operation near the critical operating point. Design and operational strategies for avoiding critical point operation can be illustrated with the aid of Equation 3.31 and the use of critical point curves.

Critical Point Curves

A critical point curve is essentially a graphical technique that quantifies the critical operating point, i.e., the values of the engineering constants at which the reactor attains a growth rate equal to μ^*, the peak growth rate given by the inhibition function. These curves can be generated using Equation 3.31. For example, critical detention time t^* is plotted versus influent substrate concentration, S_i. Consider the graph depicted in Figure 3.5. The solid line represents the values of $\bar{t}$ that result in a growth rate of μ^* for specified values of the biokinetic constants, the engineering control variables, and X_R. The cross-hatched area above the curve comprises a safe operating region where the reactor is not apt to experience sudden effluent deterioration and washout. Conversely, in the area below the curve in Figure 3.5, reactor substrate would be higher than S^* because the reactor would be operated on the downward side of the inhibition function, which "exceeds" μ^*. Operation well above the curve is necessary in order to prevent failure.

The use of critical point curves to avoid design or operation near μ^* for activated sludge reactors treating inhibitory wastes is demonstrated using

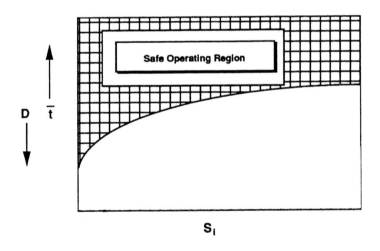

Figure 3.5.   **Critical point curve for activated sludge reactor treating an inhibitory waste. The curve is defined by substitution of appropriate biokinetic constants and selected values of α and X_R into Eq. 3.31.**

the curves presented in Figure 3.6. These curves were constructed by utilizing values of the biokinetic constants for phenol. The values of these constants are the same as those used to generate the curves shown in Figure 3.2. For Figure 3.6, Equation 3.31 was used, and it was assumed that substrate concentration in the recycle flow is negligible, i.e., $S_R = 0$. All curves were generated using a recycle ratio of 0.25; the recycle sludge concentration X_R varied from 6000 to 30,000 mg/L. It can be seen that increasing X_R values decrease the detention time required to prevent reactor failure due to μ^* violations. The reason for this trend is simply that an increase in X_R will reduce specific growth rate in the

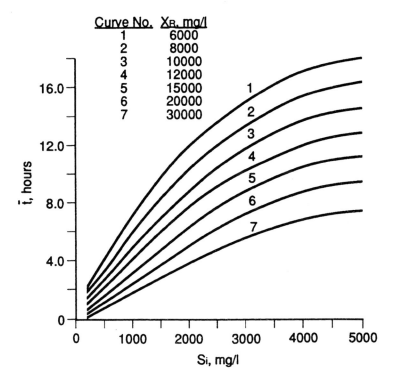

Figure 3.6. **Critical point curves generated using Equation 3.31 showing effects of X_R on location of critical operating point. Biokinetic constants are the same as those for Figure 3.2. $\alpha = 0.25$ for all curves. (Adapted from Rozich and Gaudy, 1984).**

reactor (increase Θ_c), provided the other engineering and biokinetic constants remain relatively unchanged.

It should be also noted that Figure 3.6 depicts various operational options that can be implemented in order to avoid a reactor washout. As an example of the utility of critical point curves, consider Figure 3.6 and an activated sludge reactor operating at an X_R of 6000 mg/L, α of 0.25, S_i of 1000 mg/L, and detention time of 10 hr. If it were proposed to increase the influent waste concentration S_i to 2000 mg/L, or expected that in time it could reach this concentration, then the system would be thrust beyond the critical operating point and would inevitably experience sudden washout. However, a number of options are available that can be implemented to prevent reactor failure. For example, a design decision to increase reactor detention time to approximately 13 hr would avoid operation in the proximity of the critical point. It can be seen in Figure 3.6 that this change would place the reactor operating point above the 6000 mg/L X_R line, which comprises a safe operating region. A safe operating condition can also be provided by manipulating the recycle sludge concentration, X_R, while maintaining reactor detention time at 10 hr. If X_R is increased to 10,000 mg/L, then the critical operating point curve will be shifted to the 10,000-mg/L line. At this X_R and $\bar{t}$ of 10 hr, the reactor will be in a safe operating region because it will be above the critical point curve.

It should be noted that the recycle flow ratio α can also be used for averting critical operating conditions in the aeration tank. Consider the set of critical point curves depicted in Figure 3.7. These plots illustrate the effect of α on the critical curve for this reactor system. Recalling the previous example depicted in Figure 3.6, it can be seen that at an α of 0.25, an X_R of 6000 mg/L, a detention time of 10 hr, and an S_i of 1000 mg/L, the reactor is in a safe operating region above the curve. However, if S_i changes to 2000 mg/L, the system is placed below the critical curve, meaning that the system is vulnerable to sudden effluent deterioration. Figure 3.7 depicts a number of engineering changes that could be implemented to alleviate a washout condition. For example, a change in α from 0.25 to 0.50 would realize a safe operating condition in the reactor. Also, provision of more aeration volume, i.e., an increase in reactor detention time to about 13 hr, would put the system above the critical point curve. Thus, a change in α, X_R, or $\bar{t}$ can be used to prevent failure under the new imposed loading condition.

Once a treatment plant is on line, the only manipulative alternatives available to the operator are changes in recycle sludge concentration X_R and recycle flow ratio α, provided these alternatives have been made

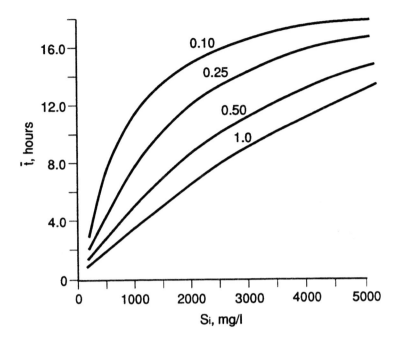

Figure 3.7. Critical point curve showing effects of a on location of critical operating point. $X_R = 6000$ mg/L, $\alpha = 0.10, 0.25,$ 0.50, 1.0. Same biokinetic constants as Figure 3.6. (Adapted from Rozich and Gaudy, 1984).

available by the designer. There may be a tendency in design to obtain the μ_n or Θ_c needed to produce the required effluent substrate concentration by holding the reactor volume, and thus $\bar{t}$, at a low value, depending on the attainment of high X_R or α, or both. For example, for an S_i of 1000 mg/L, if one wished to employ a 3-hr detention time and an α of 0.25, the recycle sludge concentration to avoid washout would be 20,000 mg/L according to the example in Figure 3.6; this would just keep the process above the critical curve and thickening would probably be required. Conversely, if one took a somewhat conservative estimate of the concentration of X_R that could be depended upon due to compaction in the clarifier, i.e., 6000 mg/L, α would need to be increased above 1.0 to avoid failure of the system. However, recycle rates above 1.0 impose a higher solids flux loading to the secondary clarifier, which means that underflow concentrations of 6000 mg/L might not even be attainable. It is thus argued (especially in the treatment of toxic or inhibitory wastes where the plant needs to be protected against sudden failure, which occurs as the reactor approaches μ^*) that it is wise to allow ample detention time or reactor volume. Such a provision, coupled with provision for flexibility in recycle pumping capacity and measures for positive control of X_R, give the operator the ability to keep the process on line.

It should be emphasized that manipulations of X_R, α, and $\bar{t}$ discussed here are related to the washout or failure curves for the plant. That is, the analysis presented herein is related to the critical reactor condition for μ and S. For the example used here, the critical condition corresponds to $\mu^* = 0.07$ hr^{-1}, S$^* = 55$ mg/L, or $\mu_n{}^* = 1.2$ d^{-1}, $\Theta_c = 0.83$ d. These are much higher values of μ_n and S than those for which the treatment plant would be designed, so that an average plant condition should be well above the critical curve. However, when one takes into consideration uncertainty factors such as possible changes in the numerical value of the biokinetic constants, inflow rate, and waste strength, and the general lack of control of α and X_R, there is ample reason to incorporate as much of a safety factor as possible. In design, this usually means the use of additional hydraulic detention time. This seems especially important for the case of inhibitory wastes as compared to noninhibitory wastes. In the latter case, an occasional period of increase in soluble substrate in the effluent might be tolerated or a fine paid for a permit limit excursion, but with inhibitory or toxic wastes, the consequences can be total failure or washout of the system, which would result in an extended period of permit violation and corresponding fines. The analysis and the critical curves presented here can be looked upon as a form of "reliability analysis" to prevent total process failure, which would require a shutdown and restart of the plant.

KEY CONCEPT SUMMARY

The key concepts contained in this chapter are:

- The derivation of predictive equations for activated sludge processes involves a mass balance approach. Mass balances for biomass and substrate are written around the reactor. An appropriate growth rate expression (Monod or Haldane) is used to relate μ to S. The equations are then solved to provide predictive equations for effluent quality S and reactor biomass concentration X.

- The modeling procedure described in this chapter can be applied to any reactor configuration to yield configuration-specific predictive equations.
- The operational location of a reactor's critical point, i.e., the point at which an activated sludge reactor treating an inhibitory waste attains the peak growth rate, μ^*, is quantifiable in terms of the engineering control parameters and the influent waste concentration. The engineering controls are then used to avoid operation near the critical point and provide input regarding a safe operating cushion using the process model.

REFERENCES AND SUGGESTED ADDITIONAL READING

Gaudy, A.F. Jr., and Gaudy, E.T. (1988). *Elements of Bioenvironmental Engineering*. Engineering Press, Inc., San Jose, California.

Rozich, A.F., Gaudy, A.F. Jr., and D'Adamo, P.C. (1983) "Predictive Model for Treatment of Phenolic Wastes by Activated Sludge." *Water Research, 17*, 1453-1466.

Rozich, A.F., and Gaudy, A.F. Jr. (1984). "Critical Point Analysis for Toxic Waste Treatment," *J. Environ. Eng. Div., ASCE, 110*, pp. 562-572.

Rozich, A.F., and Gaudy, A.F., Jr. (1985). Response of Phenol Acclimated Sludge to Quantitative Shock Loading, *J. Water Poll. Control Fed., 57*, pp. 795-804.

Rozich, A.F., Gaudy, A.F., Jr., and D'Adamo, P.C. (1985). "Selection of Growth Rate Model for Activated Sludges Treating Phenol," *Water Research, 19*, pp. 481-490.

Rozich, A.F. and Gaudy, A.F. Jr. (1986) "Process Technology for the Biological Treatment of Toxic Organic Wastes," *Hazardous and Industrial Waste Testing and Disposal, 6*, ASTM STP 933, pp. 319-333.

4 COMPARISON WITH OTHER APPROACHES AND BASIC APPLICATIONS

INTRODUCTION

The purpose of this chapter is to compare and to reconcile the activated sludge modeling approach presented herein with other more "conventional" methods or parameters often used to analyze these systems. This chapter also illustrates some basic applications concerning the use of the modeling approach to understand and to analyze various ancillary features that are often associated with the design and operation of activated sludge systems, such as:

- Impact of recycle parameters on system growth rate.
- Determination of conditions needed for minimizing sludge production.
- Computation of oxygen transfer requirements.

The objective of this chapter is to demonstrate to the reader that the modeling approach presented herein (mostly found in Chapter 3) will provide analysts of activated sludge reactors with essentially the same results as the conventional methods. However, because this approach contains a higher degree of flexibility, it enables environmental technologists to identify the limits of the system operating envelope more accurately, which enables them to "fine tune" the design and operation of activated sludge systems with a more structured, quantitative approach.

RECONCILIATION WITH OTHER DESIGN AND OPERATIONAL APPROACHES

The design and operational methodology presented here represents methodology that is not often used to analyze activated sludge systems. Nevertheless, while the approach is different, it is prudent engineering practice to evaluate how it compares with approaches that employ "food-to-microorganism" (F:M) ratio, substrate utilization rate U, and mean cell residence time Θ_c (which is simply the reciprocal of the net growth rate μ_n). F:M and U are computed using Equations 4.1 and 4.2:

$$F:M = S_i/X\bar{t} \tag{4.1}$$

$$U = (S_i - S)/X\bar{t} \tag{4.2}$$

Inspection of these equations shows that these two parameters are essentially the same, since most designs aim for low effluent substrate, or waste, concentrations where $S_i >> S$.

Mean cell residence time or Θ_c is defined by the mass of cells in the system divided by the rate of cell wasting. It is computed using Equation 4.3:

$$\theta_c = VX/X_w \tag{4.3}$$

However, the amount of biomass wastage, X_w, is also defined as the amount in excess of the amount needed for recycle:

$$X_w = (1 + \alpha)FX - \alpha FX_R \tag{3.15}$$

This equation can be rewritten similar to equation 3.17:

$$X_w = (1 + \alpha - \alpha X_R/X)VX/\bar{t} \tag{3.17}$$

From Equations 3.18 and 4.3, it is known that:

$$\mu_n = X_w/VX = 1/\theta_c \tag{4.4}$$

It is thus clear that:

$$\mu_n = X_w/VX = (1 + \alpha - \alpha X_R/X)/\bar{t} = 1/\theta_c \tag{4.5}$$

Equation 4.5 makes several important points. First, it analytically demonstrates that Θ_c and μ_n are reciprocals of one another; controlling one controls the other. Secondly, it shows that there are two ways in which Θ_c or μ_n can be calculated: (1) a feed forward mode that utilizes the recycle parameters or (2) a resultant mode using the wasting rate. In a steady-state situation, both methods will produce the same result. Finally, Equation 4.5 shows that the recycle parameters, α and X_R, exert a strong influence over Θ_c and μ_n; this becomes extremely important especially when considering the treatment of toxics or difficult-to-degrade wastes.

By using Equation 3.6 from Chapter 3 and the definition of U as being the rate of waste utilization ($F(S_i - (1 + \alpha)S$) divided by the system biomass, VX, it can be shown that:

$$\mu = Y_t U \qquad (4.6)$$

Using Equation 4.6 and Equation 3.4 from Chapter 3, it can also be shown that:

$$\mu_n = Y_t U - k_d = 1/\theta_c \qquad (4.7)$$

Equations 4.5 and 4.7 quantify the relationship between the engineering control parameters, recycle sludge flow ratio α and concentration X_R, and the more traditional parameters of F:M, U, and Θ_c. It should thus be obvious that the traditional parameters represent aggregate system indicators that result from the interactions of several factors; the same can be stated for μ_n. The alternate approach as presented here isolates the germane engineering and biokinetic parameters and determines the resulting aggregate impact on system performance using the process model. Thus, the engineer can quantify or determine which values of α or X_R, etc., are needed to meet target μ_n (or F:M or Θ_c, if those terms are preferable) values for a specific set of influent conditions and values of the biokinetic constants.

As a further illustration, consider the curves presented in Figure 4.1. Figure 4.1 shows the impact of the engineering parameters on resulting values of F:M, Θ_c, and μ_n. Predicted calculations for S were made using Equation 3.10, while Equations 4.1 and 4.7 were used to compute F:M and Θ_c, respectively; values of X, the reactor biomass concentration, were computed using Equation 3.14. These calculations were made using an S_i of 250 mg/L and values of the biokinetic constants of 0.5 hr^{-1}, 100 mg/L, 0.60, and 0.04 d^{-1}, for μ_{max}, K_s, Y_t, and k_d, respectively.

Figure 4.1 shows that there are several possible combinations of α, X_R, Y_t, and $\bar{t}$ that provide the required value of S. Thus, even though an engineer may specify one value of F:M or Θ_c, there are several ways in

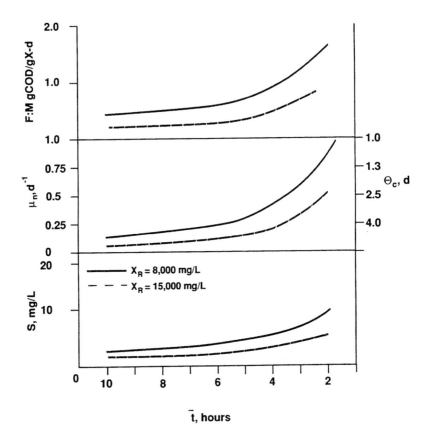

Figure 4.1.   Effect of X_R and t on S (μ_{max} = 0.5 hr^{-1}, K$_s$ = 100 mg/L, Y$_t$ = 0.6, k$_d$ = 0.04 d^{-1}). S$_i$ = 250 mg/L, α = 0.25.

which one can obtain these target values. It is also interesting to note that a recycle flow ratio α of 0.25, a recycle sludge concentration of 8,000 mg/L, and a detention time of 8 hr result in a Θ_c of 5 d. Empirically, these values are frequently cited as the values for the recycle parameters and tank detention time that are needed in order to achieve good process performance for municipal wastes. A key point is that influent waste concentrations are often greater than 250 mg/L usable COD. When this occurs, the empirical "rules of thumb" break down, since they are based on the assumption of relatively low influent concentrations for municipal wastes.

Consider Figure 4.2. All engineering parameters and biokinetic constants are the same with the exception of S_i, which is now 1000 mg/L (instead of 250 mg/L as it was in Figure 4.1). The impact of influent waste strength on the prediction of effluent quality is significant as evidenced by the substantial amount of substrate leakage predicted by the model. Furthermore, the conventional approach of utilizing F:M or Θ_c calculations does not provide the ability to determine which combinations of values for the engineering parameters are suitable for insuring good treatment. That is, the modeling approach presented herein enables the process analyst to gauge the performance envelope. Stated another way, the modeling approach allows the engineer to determine what opti-

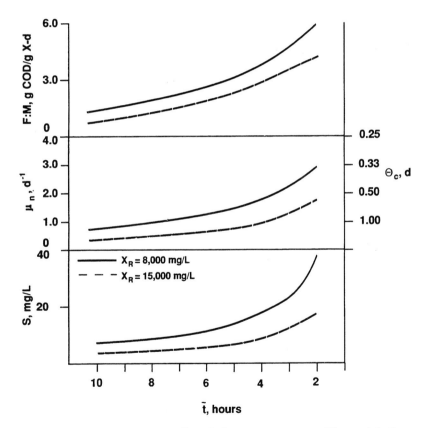

Figure 4.2.   Effect of X_R and $\bar{t}$ on S. Same constants as Figure 4.1. S_i = 1000 mg/L and α = 0.25. Compare with Figure 4.1 and note effect of increased S_i.

mal combination of the engineering parameters are appropriate for achieving target F:M or Θ_c values at higher influent waste concentrations.

IMPACT OF RECYCLE PARAMETERS ON SYSTEM GROWTH RATE

The importance of this section is that it points out that the recycle parameters, α and X_R, have a significant impact on the reactor growth rate. It will also be demonstrated how Equation 4.5 can be used to explain qualitatively various aspects of activated sludge system kinetics. Equation 4.5 is given again below:

$$\mu_n = (1 + \alpha - \alpha X_R/X)/\bar{t} = 1/\theta_c \qquad (4.5)$$

This equation underscores the advantages of maintaining high recycle sludge concentrations. If, for example, a plant has a settling or clarifier problem and the recycle sludge concentration is essentially the same as the mixed liquor concentration (i.e., $X_R \approx X$), Equation 4.5 shows that μ_n (Θ_c) is now heavily dependent on $\bar{t}$ and that the effect of the recycle parameters are essentially inconsequential. Many times plants tend to use high recycle ratios (1 or greater). This practice is not only costly from an energy (pumping costs) perspective, but it has the other negative effect of increasing solids flux (and/or surface loading) rates to the secondary clarifier. Operators and designers should thus be fully cognizant that both α and X_R can be used to control growth rate. Control of recycle sludge concentration can be implemented in the clarifier or can be achieved by using thickened waste activated sludge as a source to increase the concentration of recycle sludge. The point is that control of recycle sludge concentration is not very difficult to engineer, especially in view of the level of control it imparts for operation of an activated sludge system.

Bulking is a problem which has impacted many treatment plants. These situations are the result of changes in the ecology of the biomass, leading to a sludge with poor settling, compaction, and solids separation characteristics. Many times bulking is caused by the proliferation of filamentous microorganisms, which produce a biomass with string-like growths protruding from the sludge flocs. A filamentous bulking situation presumably occurs due to changes in environmental conditions (pH, waste characteristics, etc.) or by naturally occurring fluctuations in the system ecology. Equation 4.6 is useful for explaining in a qualitative sense why the performance of an activated sludge system is apt to deteriorate so quickly once this type of settling problem begins. Consider that the first effect of a bulking problem is a decrease in the underflow, or recycle, sludge concentration. As X_R drops, Equation 4.6 shows that μ_n increases; so does the rate of filament growth, which in turn results in increased filament proliferation. This intensifies the level of bulking and presumably causes further drops in X_R; further drops in X_R simply act to exacerbate the situation, since growth rate continues to increase. This leads to suspended solids losses over the secondary clarifier weirs and, if the situation is severe enough, excessive increases in system growth rate, as indicated by Equation 4.6, suggests the potential for leakage of soluble substrate (COD).

As discussed in Chapter 3, the treatment of inhibitory wastes mandates the avoidance of operation near the peak growth rate μ^* for the system. Equation 4.6 indicates that fluctuations in X_R produce fluctuations in growth rate which, for inhibitory waste treatment situations, can represent violations of μ^* with resulting consequences being severe effluent deterioration and potential system failure. Unless other provisions are made to provide a growth rate cushion (e.g., by using large tank volumes to allow for long hydraulic detention times), some means to control recycle sludge concentration are necessary in order to increase the performance envelope of activated sludge systems treating toxic or inhibitory wastes.

DETERMINATION OF CONDITIONS NEEDED FOR MINIMIZING SLUDGE PRODUCTION

Extended aeration plants are generally thought of as activated sludge systems that are designed and operated with the goal of having little or no excess sludge production (that is, $X_w = 0$). The question, then, is how to select the engineering control parameters to approach the condition of zero sludge wastage, or to try to minimize sludge production. One approach is to simply "waste less and see what happens." Alternatively, one can use the modeling approach presented herein and proactively determine a realistic sludge minimization goal for a given treatment facility.

First, it is important to recognize that the condition of $X_w = 0$ (zero sludge wastage) also means that the net growth rate, μ_n, is also equal to 0.

This, in turn, means that $\mu = k_d$ since $\mu_n = \mu - k_d$. Once again, recall Equation 4.5:

$$\mu_n = (1 + \alpha - \alpha X_R/X)/\bar{t} = 1/\theta_c \qquad (4.5)$$

For $\mu_n = 0$, the term in parentheses must equal 0. For this to occur, the requisite reactor biomass concentration X that will prevail under zero sludge production conditions is determined as follows:

$$X = X_R (\alpha/(1 + \alpha)) \qquad (4.8)$$

Effluent quality, S, is determined by recognizing that $\mu = k_d$:

$$\mu = k_d = \mu_{max}S/(K_s + S) \qquad (4.8a)$$
$$S = k_d K_s/(\mu_{max} - k_d) \qquad (4.8b)$$

Combining Equations 3.6 and 4.5 produces Equation 4.9, which calculates the detention time $\bar{t}_e$ required to achieve extended aeration conditions for specified values of the engineering parameters α, X_R, and S_i:

$$\bar{t}_e = \frac{1}{k_d} \left[\frac{Y_t (S_i - (1 + \alpha)S) + \alpha X_R}{X_R (\alpha/(1 + \alpha))} - (1 + \alpha) \right] \qquad (4.9)$$

The values of the biokinetic constants impact the prediction of sludge minimization conditions by virtue of the fact that the prediction of effluent quality, S, is given by Equation 2.11. For this application, $\mu_n = D = 0$. Equation 4.9 is simplified to Equation 4.9a:

$$\bar{t}_e = \frac{1}{k_d} \left[\frac{Y_t (S_i - (1 + \alpha)S) + \alpha X_R}{\alpha X_R/(1 + \alpha)} - (1 + \alpha) \right] \qquad (4.9a)$$

Now consider a treatment plant with a recycle sludge concentration of 10,000 mg/L, a recycle flow ratio of 0.25, and biokinetic constant of 3.15 d^{-1}, 0.07 d^{-1}, 105 mg/L, and 0.63 for μ_{max}, k_d, K_s, and Y_t, respectively. Table 4.1 presents the calculated required detention times for achieving extended aeration ($\mu_n \approx 0$) conditions for a variety of values of S_i, the influent waste strength.

Table 4.1 makes some interesting points. First, it is useful to recall the old axiom that extended aeration conditions are achieved with a 24-hr detention time. Consider the fact that "typical" operating parameters for municipal treatment facilities are the recycle values used in the table

Table 4.1 Required Detention Time for Achieving Extended Aeration Conditions in an Activated Sludge System

S_i (mg/L COD)	t_e (hours)*
100	11
250	27
500	54
1,000	108

*Computed using Equation 4.9a.

calculations (α of 0.25 and X_R of 10,000 mg/L) and that primary effluents for municipal systems are generally close to 250 mg/L COD. The model shows that, with these conditions and specified values of the biokinetic constants, a plant will achieve extended aeration at a detention time of 27 hr. This value is relatively close to the 24-hr detention time cited so often.

As S_i increases, the required detention time also increases. This occurs because S_i has a profound influence on reactor growth rate. In order to keep the growth rate at the required value ($\mu = k_d$) for minimizing sludge production, it is necessary to increase $\bar{t}$ substantially. The key point is that this model analysis enables process analysts to determine the potential for minimizing sludge production at their facilities. The model provides the required $\bar{t}$ and the expected mixed liquor suspended solids values (X) for various operating conditions (α, X_R, and S_i). Thus, facility operators who desire to minimize sludge production in their systems can do so using a proactive strategy that enables them to make the assessment using a structured algorithm as a management tool.

COMPUTATION OF AERATION TANK OXYGEN TRANSFER REQUIREMENTS

Another useful application of the modeling approach presented here is with regards to the computation of oxygen transfer requirements for activated sludge reactors. Many times, design or operation problems involve the detailed consideration of oxygen transfer. This is not a trivial consideration; the optimization of the aeration facilities can represent substantial cost savings since the operation of these systems requires a relatively large power commitment. The approach presented herein is targeted for both design and operational applications. That is, this approach can be used to size transfer facilities for new systems as well as to optimize or modify the operation of existing ones.

The modeling methodology for predicting oxygen transfer requirements integrates our modeling concepts with the fact that all metabolized COD in an aerobic reactor is channelled into either oxygen uptake (CO_2 evolution) or into cell COD:

$$\Delta COD = O_2 \text{ uptake } + \Delta COD_{cells} \qquad (1.17)$$

This balance concept, which was presented in detail in Chapter 1, and the modeling equations presented in Chapter 3, are combined to provide predictive equations for the required k_{la} for a reactor.

As an illustration of the application, consider a simple chemostat as depicted in Figure 4.3. Steady-state mass balances yield predictive equations for X and S, the effluent substrate and biomass concentrations, respectively:

$$t_e = \frac{(1 + \alpha)}{k_d} \frac{Y_t (S_i - (1 + \alpha)S)}{\alpha X_R} \qquad (4.10)$$

$$S = \frac{K_s (D + k_d)}{\mu_{max} - (D + k_d)} \qquad (4.11)$$

$$X = \frac{Y_t D(S_i - S)}{D + k_d} \qquad (4.12)$$

An additional mass balance is also written for dissolved oxygen concentration, C:

$$
\overbrace{V \frac{dC}{dt}}^{\substack{\text{rate of } O_2 \\ \text{change}}} = \overbrace{FC_i + (k_{la} (C_s - C)V)}^{\substack{\text{rate of } O_2 \\ \text{transfer}}} - \overbrace{((F)(C) - O_2 \text{ for metabolic demand})}^{\substack{\text{rate of } O_2 \text{ leaving} \\ \text{reactor}}}
$$
$$(4.13)$$

The oxygen uptake rate for the chemostat is computed using the aerobic balance principle shown in Equation 4.14.

$$O_2 \text{ uptake } = F(S_i - S) - 1.42FX \qquad (4.14)$$

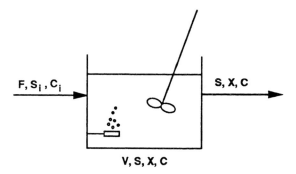

Figure 4.3.   Chemostat mass balances for cells, substrate, and oxygen.

It should be noted that, for the chemostat, $F(S_i - S)$ represents the ΔCOD and $1.42FX$ represents the ΔCOD_{cells}. Inserting Equation 4.14 into Equation 4.13 results in Equation 4.15:

$$V \frac{dC}{dt} = k_{la}(C_s - C)V - FC - (F(S_i - S) - 1.42XF) \qquad (4.15)$$

Expressions for S and X are given in Equations 4.11 and 4.12, respectively. The steady-state solution for predicting k_{la} in a chemostat is given in Equation 4.16.

$$k_{la} = \frac{(C + S_i - S - 1.42X)}{\bar{t}(C_s - C)} \qquad (4.16)$$

Equation 4.16 is a function of the engineering parameters $\bar{t}$ and S_i, the biokinetic constants, the reactor dissolved oxygen concentration (DO), and the saturation DO value. This equation gives the required k_{la} for aeration equipment for a specified set of conditions in a chemostat.

The application of this approach for obtaining a predictive equation for k_{la} in a completely mixed activated sludge reactor uses Figure 4.4 and a similar analytical approach. The oxygen uptake rate in a completely mixed activated sludge system is provided by Equation 4.17:

$$\overset{\Delta COD}{} \qquad\qquad \overset{\Delta COD_{cells}}{}$$

$$O_2 \text{ uptake} = (F(S_i + \alpha S_R) - (1 + \alpha)FS) - ((1 + \alpha)FX - \alpha FX_R)1.42)$$
$$(4.17)$$

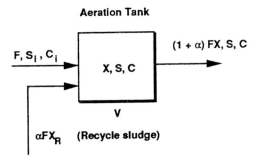

Figure 4.4. One cell completely mixed activated sludge mass balances for cells, substrate, and oxygen.

Predictive equations for S and X were derived and are given in Chapter 3; these are Equations 3.10 if the waste is noninhibitory or 3.19 if the waste is inhibitory (for S) and Equation 3.14 (for X). The oxygen mass balance is given below:

$$V \frac{dC}{dt} = k_{la}(C_s - C)V - (1 + \alpha)FC \qquad (4.18)$$

$$- F(S_i + \alpha S_R - (1 + \alpha)S - 1.42((1 + \alpha)X - \alpha X_R))$$

A steady-state predictive equation for k_{la} in a completely mixed activated sludge reactor is given in Equation 4.19:

$$k_{la} = \frac{F((S_i + \alpha S_R - (1 + \alpha)S - 1.42((1 + \alpha)X - \alpha X_R)) + (1 + \alpha)C)}{(C_s - C)V} \qquad (4.19)$$

A situation which is perhaps of more practical importance from both a design and operational perspective is the prediction of the variation of required oxygen transfer capacity or k_{la} in a plug flow reactor. The same modeling approach employed to obtain the multiple-cell predictive model in Chapter 3 is utilized to derive the predictive equations for k_{la} in a multiple reactor, or plug flow system. The difference here is that we incorporate the aerobic balance principle, perform an oxygen mass balance in each cell, and use Figure 4.5. In the multiple-cell system, one will predict different k_{la} values for each cell. This enables designers to refine equipment selection or managers to refine operating policies to conserve energy expenditures. For the derivation, it is assumed that there are "m"

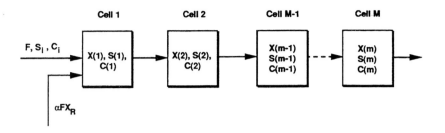

Figure 4.5. Multiple-cell activated sludge mass balances for cells, substrate, and oxygen.

number of completely mixed aeration tanks of equal volume. If the total tank volume is V, then each cell has a volume of V/m.

An expression for the oxygen uptake rate in the first tank in Figure 4.4 is provided in Equation 4.20:

$$\overbrace{\phantom{(F(S_i + \alpha S_R) - (1 + \alpha)FS_{(1)})}}^{\Delta COD} \quad \overbrace{\phantom{(((1 + \alpha)FX_{(1)} - \alpha FX_R)1.42)}}^{\Delta COD_{cells}}$$

$$O_2 \text{ uptake} = (F(S_i + \alpha S_R) - (1 + \alpha)FS_{(1)}) - (((1 + \alpha)FX_{(1)} - \alpha FX_R)1.42) \quad (4.20)$$

The oxygen mass balance in tank 1 is:

$$\frac{V}{m}\frac{dC_{(1)}}{dt} = k_{la}(C_s - C_{(1)})\frac{V}{m}$$

$$- (1 + \alpha)FC_{(1)} - F(S_i + \alpha S_R - (1 + \alpha)S_{(1)} - 1.42((1 + \alpha)X_{(1)} - \alpha X_R)) \quad (4.21)$$

At steady state, the predictive equation for k_{la} in tank 1 is:

$$k_{la_1} = \frac{mF((S_i + \alpha S_R - (1 + \alpha)S_{(1)} - 1.42((1 + \alpha)X_{(1)} - \alpha X_R)) + (1 + \alpha)C_{(1)})}{(C_s - C_{(1)})V} \quad (4.22)$$

Equations for $X_{(1)}$ and $S_{(1)}$ are given in Chapter 3.

For tanks 2 through m, the expression for oxygen uptake rate is:

$$\overbrace{\phantom{(1 + \alpha)F((S_{(j-1)} - S_{(j)})}}^{\Delta COD} \quad \overbrace{\phantom{(1.42(X_{(j)} - X_{(j-1)})))}}^{\Delta COD_{cells}}$$

$$O_2 \text{ uptake} = (1 + \alpha)F((S_{(j-1)} - S_{(j)}) - (1.42(X_{(j)} - X_{(j-1)}))) \quad (4.23)$$
tanks 2
through m

The oxygen mass balance equation for tanks 2 through m is:

$$\frac{V}{m} \frac{dC_{(j)}}{dt} = k_{la}(C_s - C_{(j)}) \frac{V}{m}$$

$$- (1 + \alpha)FC_{(j)} - (1 + \alpha)F(S_{(j-1)} - S_{(j)} - 1.42(X_{(j)} - X_{(j-1)})) \quad (4.24)$$

At steady state, the predictive equation for k_{la} for tanks 2 through m in a multiple-cell or plug flow reactor is given in Equation 4.25.

$$k_{la} = \frac{m(1 + \alpha)F(S_{(j-1)} - S_{(j)} - 1.42(X_{(j)} - X_{(j-1)})) + C_{(j)}}{(C_s - C_{(j)})V} \quad (4.25)$$

It should be noted that the equations presented in this section can be used to predict steady-state oxygen uptake requirements in either completely mixed or multiple-cell reactor configurations. The process analyst can examine numerous different design or operational scenarios for oxygen transfer requirements by simply changing the various input parameters. These parameters are the recycle variables, α and X_R, the influent flow rate and waste concentration, F and S_i, respectively, and the biokinetic constants. One could also investigate the optimum number of reactor cells that can be utilized to minimize the aggregate tank k_{la} requirements. This approach provides the activated sludge system designer, operator, and manager with a structured and relatively straightforward algorithm for quantifying and predicting oxygen transfer requirements. The use of this tool will enable process analysts to refine the optimization of the oxygen transfer features for activated sludge system design and operation.

KEY CONCEPT SUMMARY

The key concepts contained in this chapter are:

- The approach to activated sludge presented herein is readily reconcilable with other, more conventional approaches, i.e., F:M, U, and Θ_c. The modeling approach presented in this text differs in that it is more flexible and that it enables the process analyst to examine the impact of the individual process components on system performance (waste strength, recycle parameters, etc.). This enables the process analyst to define the performance envelope more accurately.

- The recycle parameters for an activated sludge system greatly impact system growth rate and thus significantly influence process performance. The approach presented here can be used to quantify the optimal values of these parameters. This is especially crucial for systems with high loadings or those that must handle inhibitory or difficult-to-degrade wastes.
- The determination of the specific conditions, i.e., the values of the recycle parameters, tank detention times, etc., needed to minimize waste sludge production can be quantified.
- Predictive equations for determining aeration requirements (k_{la} values) for activated sludge systems can be formulated with a relatively simple application of the process model. This enables process analysts to refine design or operational issues that involve aeration equipment.

REFERENCES AND SUGGESTED
ADDITIONAL READING

Gaudy, A.F. Jr., and Gaudy, E.T. (1988). *Elements of Bioenvironmental Engineering,* Engineering Press, Inc., San Jose, California.

Gaudy, A.F. Jr., Srinivasaraghavan, R., and Saleh, M. (1977). "Conceptual Model for Design and Operation of Activated Sludge Processes." *J. Environ. Eng. Div., ASCE, 103*, pp. 71–84.

5 PROCEDURES FOR OBTAINING BIOKINETIC CONSTANTS

INTRODUCTION

This chapter will detail analytical and laboratory methodologies that can be used to generate and then analyze the requisite data needed to quantify the values of the biokinetic constants. It is the goal of this chapter to provide a "cookbook" approach that will enable practitioners to determine the values of these parameters without having to initiate a major "research" effort. That is, our experience has been that determination of the constants can be performed on a routine basis without much difficulty, provided that a reasonably experienced technician is available for performing the work.

This chapter contains three main sections:

- Determination of Influent Waste Strength, S_i
- Use of Respirometry to Generate Kinetic Data
- Determination of True Cell Yield, Y_t, and Decay Rate, k_d

The determination of influent waste strength, S_i, is not a trivial issue. This is stated because waste streams are rarely, if ever, composed primarily of one or only a few components. Consequently, the delineation of what "S_i" is for a biological treatment system can at times be a difficult task. This chapter will address this issue and present techniques and strategies that can be employed to quantify influent waste strength.

The biokinetic growth constants, which are contained in the growth rate expressions which relate μ to S (biomass growth rate to substrate or waste concentration), are the biological parameters having the most influence on predicting effluent quality. Also, these parameters are the ones more likely to exhibit the greatest level of variation with changing environmental conditions. As previously discussed, there are three biokinetic growth constants: μ_{max}, K_s, and K_i (if the waste is inhibitory). The key data needed to determine the values of the biokinetic growth constants are growth rate vs. substrate concentration. These data are generated by analyzing respirometric or oxygen uptake measurements of the target biomass degrading the target waste; these measurements are taken in tests that can be completed in as little as eight hours and rarely go longer than one day. Once the growth rate data are collected, they are fit to the Monod or Haldane (if the kinetics are inhibitory) growth rate expressions. This chapter will detail how respirometric tests are set up, how the data are analyzed to determine growth kinetics, and then how the growth rate data are analyzed to determine the values of the biokinetic growth constants.

Finally, this chapter will discuss methods to determine values of true cell yield, Y_t, and decay rate, k_d. These constants are more crucial for predicting quantities of waste sludge production and are not as critical for predicting effluent quality. This is fortuitous, since the data needed for determining the true cell yield and decay rate must be generated in relatively long-term tests using continuous flow reactors.

DETERMINATION OF INFLUENT WASTE STRENGTH, S_i

The definition and explanation of the ΔCOD test for measurement of biologically usable COD for a particular biomass was previously discussed in Chapter 1. Its primary application regarding modeling procedures for designing and operating biological wastewater treatment systems is as a means to quantify the influent waste concentration a particular reactor must handle. It is important to note that the definition of ΔCOD comprises both a concept and test. The concept is basically contained in the fact that any given biological wastewater treatment system is characterized with a certain amount of inefficiency, which precludes the total removal of all soluble COD. That is, the soluble effluent of the biological system will always contain some COD, which many times is not associated with the influent COD since the effluent soluble COD is frequently composed of microbial excretion products and other by-products of biochemical activities. In other situations, a portion of the influent COD is resistant to biodegradation and passes through the system. Finally, a biomass may be not be acclimated to a new waste stream or component and the new carbon source, although ultimately biodegradable, may pass through the plant and appear as soluble effluent COD because the new COD may not represent a readily biodegradable carbon source for the existing biomass. (For example, consider the case of the glucose COD not being metabolized by phenol-acclimated cells, depicted in Figure 6.7). It should be noted that this conceptual consideration differs somewhat from the ΔCOD test because the test, which is described in Chapter 1, assumes that the biomass is well acclimated.

Examples of actual ΔCOD test results are given in Figures 5.1, 5.2, and 5.3. The test shown in Figure 5.1 was performed using a population that was acclimated to the readily biodegradable carbon source, glucose. In this test, the residual COD, COD_e, was quite low, and only traces of glucose were found. Thus, in this instance, the ΔCOD essentially equals the influent COD and the residual is likely attributable to microbial excretion products, intermediates, and/or components from dead or dying cells. In this case, S_i equals influent COD.

Figure 5.2 represents a quite different situation. This test was performed using an acclimated biomass on a waste with a high lignin content. It shows that the amount of COD_e is quite high, which means that

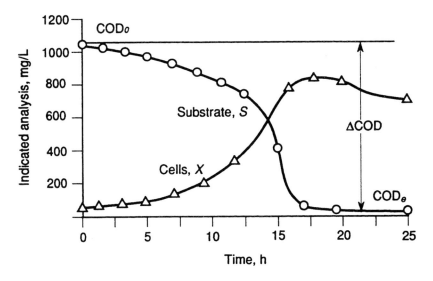

Figure 5.1. ΔCOD and biomass curves for a sample containing glucose as carbon source. (Gaudy and Gaudy, 1988).

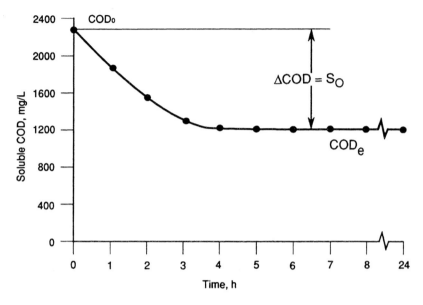

Figure 5.2. ΔCOD for a sample of dilute kraft pulp mill digestor blow-down liquor. (Gaudy and Gaudy, 1988).

there will be a relatively large difference between influent COD and ΔCOD (S_i). Lignins are not readily metabolized and are not normally expected to contribute to the organic loading of a biological wastewater treatment facility.

In Figure 5.3, the results of a ΔCOD test are presented for a municipal waste treatment plant that is treating a domestic waste with a relatively high industrial contribution. The results in Figure 5.3 depict a case that is between the ones in Figures 5.1 and 5.2. For the test in Figure 5.3, about 70% of the influent soluble COD was readily amenable to biodegradation, while the tests in Figures 5.1 and 5.2 showed results of 95% and 45%, respectively.

The use of influent COD values for determining S_i values for design and modeling purposes will provide the biological system analyst with a conservative estimate of the influent waste strength. This is stated because ΔCOD will always be less than the influent waste strength (COD_i) and the ΔCOD provides the estimate of the loading that is amenable to biodegradation by the biomass. S_i as defined using ΔCOD gives the analyst the actual organic loading that the biomass will degrade.

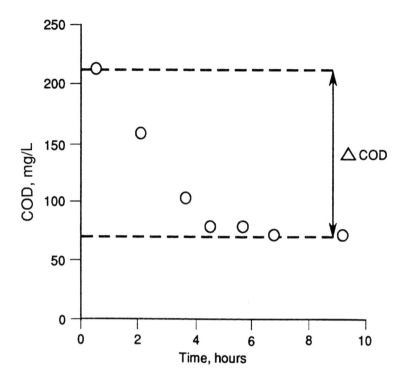

Figure 5.3.   **Example of ΔCOD test performed on municipal primary settling tank effluent. ΔCOD = 131 mg COD/L.**

It should be also be pointed out that what is COD_e for one reactor may also be COD_i or ΔCOD for another reactor. For example, consider two activated sludge reactors in series. (This approach is sometimes utilized for two-stage nitrification.) The effluent from the first system is the feed for the second system. The feed for the second system may consist of COD that passed through the first unit, metabolic intermediates, or both.

For operational purposes, the periodic determination of the ΔCOD can be used as a relatively rapid check on the plant's performance. That is, if soluble CODs in the effluent increase, it may be due to an increase in recalcitrant organics in the influent and not due to any operational oversights. It should also be noted that quantification of the S_i values via the ΔCOD procedure enables the plant to optimize operations by determining actual loading conditions, which may be different (ideally lower) than the design values. For instance, an accurate evaluation of S_i allows the plant to determine the actual number of reactors it needs in service, the level of aeration required, or both. These measures enable the plant to economize operations via the implementation of energy-saving practices.

USE OF RESPIROMETRY TO GENERATE KINETIC DATA

The key advantage of using respirometry for collecting kinetic information is the fact that data are collected much more efficiently than can be accomplished with other methods. Additionally, the data can be collected automatically and in greater quantities than is possible with alternate techniques. Previously, kinetic data for calibrating biological treatment models were collected using either the batch growth study (shake flask) method, substrate utilization technique, or various hybrids of these procedures to generate growth kinetics which can be analyzed to determine values of the biokinetic constants. Although the constants generated using these methods enabled us to obtain reasonable predictions for reactor performance, the work involved with obtaining the constants and associated analytical difficulties precluded the routine determination of biokinetic constants. The use of respirometric procedures circumvents many of these problems.

The goal with the respirometric methodology is to obtain data that can be analyzed to determine the values of the biokinetic growth constants, μ_{max}, K_s, and K_i (if the waste is inhibitory); these are the constants contained in the two growth rate relationships, Equations 1.11 and 1.12, which are utilized to relate μ to S. Once we can relate μ to S, then we have calibrated the model.

The respirometric procedure for calibrating the model, i.e., determining the values of the biokinetic growth constants, involves the generation of respirometric data which are translated into either biomass growth data or substrate utilization data. The translated data are then analyzed to obtain a set of growth rate and substrate (waste) concentration data, i.e., μ and S points, which are then fit to either the Monod or Haldane function to determine the values of the biokinetic constants μ_{max}, K_s, and K_i.

Conversion of Respirometric Data to Growth or Substrate Utilization Data

The link between oxygen uptake (respiration) and biomass growth or substrate utilization is based on the assumption that the COD being removed from solution during metabolism is channelled in varying proportions into the synthesis of new cells and to respiration measurable as oxygen uptake. This relationship between growth, substrate removal, and respiration is quantified using Equation 5.1.

$$\Delta COD = O_2 \text{ Uptake} + \Delta COD_{cells} \tag{5.1}$$

This equation states that the amount of substrate removal in an aerobic biological system is accounted for as the amount of COD that has been incorporated into the cells (ΔCOD_{cells}) plus that which has been oxidized, represented as accumulated oxygen uptake. In the absence of physical stripping or chemical oxidation of the substrate, Equation 5.1 represents the complete mass balance for substrate removal in aerobic biological systems.

Equation 5.1 can be simplified by noting that the amount of substrate COD that has been channelled to biomass (ΔCOD_{cells}) can be expressed as the product of the amount of cells produced (ΔX) and the unit COD of the cell mass (O_x). This relationship is given in Equation 5.2.

$$\Delta COD_{cells} = (\Delta X)(O_x) \tag{5.2}$$

The equation for cell yield is also used to simplify the expression.

$$Y = \Delta X / \Delta COD \tag{5.3}$$

Equation 5.3 is utilized to produce an expression for ΔCOD.

$$\Delta COD = \Delta X / Y \tag{5.4}$$

Recognizing that $\Delta X = X_t - X_0$ (X_t and X_0 are X at some time t and at time 0, respectively) and substituting Equations 5.2 and 5.4 into Equation 5.1, we obtain Equation 5.5.

$$(X_t - X_0)Y = O_2 \text{ uptake} + (X_t - X_0)O_x \tag{5.5}$$

Rearranging and simplifying Equation 5.5 yields Equation 5.6.

$$X_t = X_0 + O_2 \text{ Uptake}/(1/Y - O_x) \tag{5.6}$$

Equation 5.6 is used to convert oxygen uptake data into biomass growth curves, which can then be analyzed to determine growth rates (μ's) at different initial substrate or waste concentrations (S's); one growth curve is generated for each respirometric test unit.

In order to use Equation 5.6 for converting O_2 uptake data to biomass data, values for cell yield, Y, and cell COD, O_x, must be selected. These parameters can be determined from actual test data or estimated using previous information on the particular waste treatment system from past kinetic tests. To calculate values for Y from test data, use Equation 5.3; O_x values are calculated using Equation 5.7.

$$O_x = (Total\ COD - COD)/X \qquad (5.7)$$

O_x can be calculated during the substrate (COD) removal phase of the test and then averaged to obtain an average O_x value for use in Equation 5.6. According to Porge's stoichiometric formula for activated sludge ($C_5H_7NO_2$) (Gaudy and Gaudy, 1988), values of O_x should be close to 1.42 mg COD/mg X; in general, experimentally determined values for O_x are close to this value.

Example of Obtaining Growth Data from Respirometric Data

This example presents the results of a respirometry study performed for the Patapsco Wastewater Treatment plant (Baltimore, Maryland) (Rozich and Gaudy, Inc., 1988) using a 24-hour composite sample of primary effluent as a waste sample (S) and plant return activated sludge for biomass seed (X). Table 5.1 lists the cumulative O_2 uptake data that were collected from five reactor cells at five different initial waste concentrations. These data, plotted in Figure 5.4, were collected directly using an electrolytic respirometer. A separate batch reactor containing an undiluted waste sample and small biomass inoculum was run to collect data for calculating Y and O_x values; these data and the calculations for computing Y and O_x are given in Table 5.2. It should be noted that the data in Table 5.2 can also be used to calculate a ΔCOD value for this waste sample.

The biomass growth curves are generated by using Equation 5.6 and the data in Table 5.1, along with the values of Y and O_x computed in Table 5.2. The results of this analysis yield a computed biomass curve (X as a function of time) for the different substrate (COD) concentrations in each flask.

Table 5.1 Example of Cumulative Oxygen Uptake Data Collected from a Respirometric Study Performed on 5/12/87 Using a 24-hour Composite Sample of Primary Effluent

Cumulative Oxygen Uptake (mg O_2/L)

Time h	Percentage Waste Strength				
	20%	40%	60%	80%	100%
0	0	0	0	0	0
0.2	0	0.6	0	0.6	0
0.7	0	1.4	0.9	3.0	2.2
1.2	0	1.6	0.9	5.6	5.0
1.7	0.1	1.6	3.1	8.6	8.3
2.2	0.2	1.7	5.1	12.1	11.8
2.7	0.6	1.8	9.2	16.5	16.4
3.2	1.5	1.9	12.6	21.7	21.3
3.7	2.3	2.7	16.1	26.6	29.0
4.2	3.0	3.5	19.8	32.2	35.1
4.7	3.5	5.3	24.0	37.0	41.7
5.2	4.3	5.6	26.8	40.7	49.4
5.7	4.9	6.7	28.7	43.6	54.0
6.2	5.4	7.3	30.5	45.8	57.2
6.7	6.0	8.1	30.0	48.0	60.9
7.2	6.7	8.5	33.4	50.2	63.7
7.7	7.4	9.4	35.4	52.3	66.3
8.2	8.2	10.2	36.9	54.3	68.8
8.7	8.9	11.0	38.6	56.2	70.6
9.2	9.7	12.0	40.4	58.1	73.1
9.7	10.3	13.1	41.2	60.0	74.7
S_o (mg COD/L)	28	55	83	110	138
X_o (mg TSS/L)	40	55	80	105	145

The calculated values for X, computed using Equation 5.6, are listed in Table 5.3, while plots of the data on semilogarithmic paper are presented in Figure 5.5. The growth rate for each waste concentration is determined by identifying the exponential growth phase on each curve and drawing a straight line through the data. The slope of this line is the specific growth rate. The growth rate is determined using Equation 5.8.

$$\mu = \ln(X_2/X_1)/(t_2 - t_1) \tag{5.8}$$

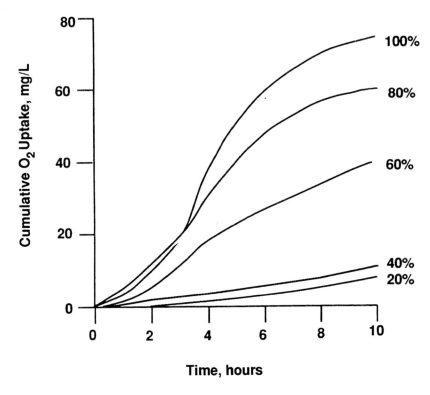

Figure 5.4.   Plot of cumulative oxygen uptake data from Table 5.1.

In Equation 5.8, X_1 and t_1 and X_2 and t_2 represent the beginning and end coordinates for the exponential phase, respectively. It should be noted that depending on if the kinetics adhere to the Monod (noninhibitory) or Haldane (inhibitory) relationship, the analyst can sometimes expect a different characteristic biomass growth curve that requires some additional consideration.

Consider the idealized growth curves presented in Figures 5.6 and 5.7. If the waste is noninhibitory, then one can expect the type of growth curve depicted in Figure 5.6 and the identification of the exponential phase and subsequent analysis to compute the growth rate are relatively straightforward. Such is not always the case for inhibitory wastes. Many times, inhibitory growth curves resemble the idealized curve shown in Figure 5.7; the specific shape of the curve is dictated by a combination of

Table 5.2 Raw Data for Determination of Y and $\bar{O}_x$

Time h	pH	Temp °C	Sample Volume ml	Suspended Solids Sample Volume ml	Tare Label	Tare Weight g	Tare + Solid Weight g	X_t mg TSS/L	Total COD mg COD/L	COD_t mg COD/L	COD of Cells O_x
0	7.0	22	35	25	8	0.3234	0.3285	204	545	239	1.50*
0.5			35	25	29	0.3406	0.3460	216	529	227	1.40
1			35	25	19	0.2802	0.2861	236	534	223	1.32
1.5			35	25	10	0.3171	0.3231	240	545	209	1.40
2		22	35	25	25	0.2909	0.2968	236	534	185	1.48
2.5			35	25	4	0.3135	0.3195	240	504	180	1.35
3			35	25	27	0.3057	0.3119	248	483	158	1.31
4			35	25	9	0.3589	0.3654	260	524	147	1.45
6			35	25	4	0.3151	0.3214	252	515	141	1.48
8.5			35	25	6	0.2983	0.3048	260	443	125	1.22*
11	8.3	22	25	25	–	–	–	–	–	127	–

$$\bar{O}_x = 1.40$$

Calculating Y:
1. Biomass growth due to soluble COD removal: 260-204 = 56 mg TSS/L
2. Soluble COD removed (ΔCOD): 239-125 = 114 mg COD/L
3. Biomass cell yield (Y): Y = 56/114 = 0.49

Calculating $\bar{O}_x$:
1. O_x (for a given sample) = (Total COD - COD_t) / X_t
2. $\bar{O}_x = \Sigma O_x$ / no. of samples: $\bar{O}_x$ = (1.40 + 1.32 + 1.40 + 1.48 + 1.35 + 1.31 + 1.45 + 1.48) / 8 = 1.40

Note: *These points were not included in the overall average.

Table 5.3 Biomass Growth Data Transformed from the Cumulative
Oxygen Uptake Data listed in Table 5.1 using Equation 5.6

Biomass Growth Data (mg TSS/L)

Time	Percentage Waste Strength				
h	20%	40%	60%	80%	100%
0	40.0	55.0	80.0	105.0	145.0
0.2	40.0	55.9	80.0	105.9	145.0
0.7	40.0	57.2	81.4	109.7	148.4
1.2	40.0	57.5	81.4	113.7	152.8
1.7	40.2	57.5	84.8	118.4	157.9
2.2	40.3	57.6	88.0	123.9	163.4
2.7	40.9	57.8	94.4	130.7	170.6
3.2	42.3	58.0	99.7	138.9	178.2
3.7	43.6	59.2	105.1	146.5	190.2
4.2	44.7	60.5	110.9	155.2	199.8
4.7	45.5	63.3	117.4	162.7	210.1
5.2	46.7	63.7	121.8	168.5	222.1
5.7	47.6	65.5	124.8	173.0	229.3
6.2	48.4	66.4	127.6	176.5	234.3
6.7	49.4	67.6	129.9	179.5	240.0
7.2	50.4	68.3	132.1	183.3	244.4
7.7	51.5	69.7	135.2	186.6	248.5
8.2	52.8	70.9	137.6	189.7	252.4
8.7	53.9	72.2	140.2	192.7	255.2
9.2	55.1	73.7	143.0	195.7	259.1
9.7	56.1	75.4	144.3	198.6	261.6
S_o (mg COD/L)	28	55	83	110	138
X_o (mg TSS/L)	40	55	80	105	145

Note: $\bar{O}_x$ = 1.40
Y = 0.49

the level of inhibition, the amount of inocula used (X_0), and the initial
COD (S) that is employed in the particular flask. For the type of curve
depicted in Figure 5.7, the correct exponential phase, which correlates
with the growth rate, is the first exponential phase that is manifested and
not the larger "second" exponential phase. Previous laboratory and analytical work on the inhibitory substrate phenol showed that analyzing the

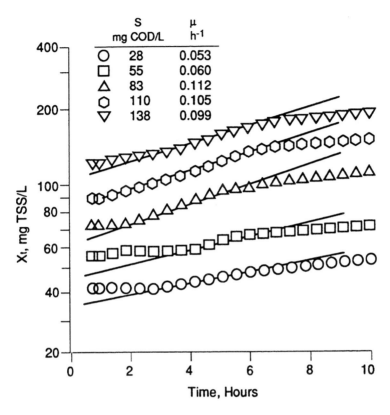

Figure 5.5. Plot of biomass growth data (see Table 5.3) at different initial waste concentrations for determination of the specific growth rate, μ.

second phase produces a large overestimate of the characteristic growth rate of the target batch reactor. This in turn will result in producing erroneously higher estimates of the biokinetic growth constants. Thus, when analyzing kinetic results for wastes that may be inhibitory, care should be exercised in determining growth rates. More detail on this aspect of analyzing growth kinetics in inhibitory systems is given elsewhere (D'Adamo, Rozich, and Gaudy, 1984).

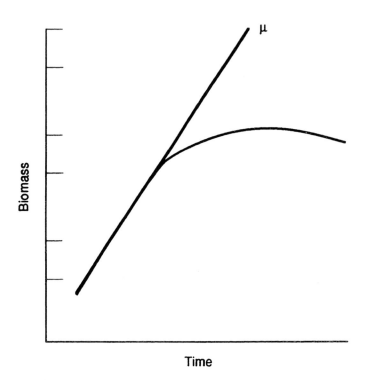

Figure 5.6. Idealized semilog plot of the increase in biomass concentration as a function of time for a noninhibitory waste. The straight-line portion of this curve represents the exponential growth phase, the slope of which is defined as the specific growth rate, μ (see Equation 1.5).

Determination of the Biokinetic Constants from Growth Data

The growth rate data (μ vs. S) obtained via analysis of the respirometric data using Equation 5.6 are fit to either the Monod or the Haldane equation to determine the values of the biokinetic constants.

$$\mu = \mu_{max}S/(K_s + S) \qquad \text{Monod} \qquad (1.11)$$

$$\mu = \mu_{max}S/(K_s + S + S^2/K_i) \qquad \text{Haldane} \qquad (1.12)$$

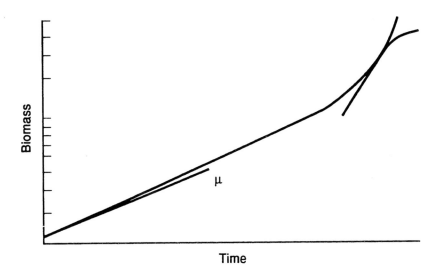

Figure 5.7. **Idealized semilog plot of the increase in biomass concentration as a function of time for an inhibitory waste. The initial straight-line portion of this curve represents the exponential growth phase, the slope of which is defined as the specific growth rate, μ (see Equation 1.5).**

If μ increases with S until it becomes asymptotic to an upper limit, the test data are described with the Monod equation (noninhibitory). Conversely, if the growth data first increase with S, reach a peak, and then begin to decrease, the Haldane equation (inhibitory) is used to fit the data.

The growth data can be fit to both the Monod and Haldane equations using a nonlinear least squares method contained in a computer program. Regarding fitting data to the Monod equation, our experience has been that nonlinear computer methods are both faster than and preferable to reciprocal plotting techniques. In the case of fitting data to the Haldane expression, reciprocal plotting techniques are virtually unusable. It was also determined that because one is fitting data to an expression with three constants, it is possible to obtain numerous (if not infinite) different combinations of μ_{max}, K_s, and K_i that all provide reasonable fits to the data. This fitting problem is addressed by incorporating a bounding procedure around the nonlinear algorithm, which provides consistent input for the initial guesses of the values of the constants. As values of the constants are generated from the nonlinear fitting procedure, they are subjected to a screening test to determine if the constant values are within the values stipulated in the bounding procedure. For example, negative values of the constants are automatically rejected. More detailed discussion on this aspect of analyzing inhibitory growth data is given elsewhere (D'Adamo, Rozich, and Gaudy, 1984).

A computer program for fitting growth data to the Monod equation is given in the Appendix; the nonlinear package, "DUNLSF," can be obtained from the International Mathematical and Statistical Libraries (IMSL) in Houston, Texas. Fitting the growth data (μ vs. S) in Table 5.3 using the program in the Appendix (values of EPS and NSIG for the DUNSLF routine equal to 0.0 and 3, respectively) yields values of 0.153 hr^{-1} and 55 mg/L COD for μ_{max} and K_s, respectively, with a residual sum of squares value (SSQ) of 0.00079. As a check, it is useful to compare a plot of the fitted curve with the actual growth data as depicted in Figure 5.8. This plot shows that although the Monod equation provided a reasonable description of the data (as also evidenced by the low SSQ value), the trend of the data can also be interpreted to suggest Haldane kinetics. Without having the luxury of additional confirmatory data (e.g., growth rate data at higher COD concentrations would be helpful), it is prudent to also fit the growth data to the Haldane expression and compare the results with the Monod fit.

A computer program for fitting growth data to the Haldane equation is given in the Appendix; it also utilizes the nonlinear fitting subroutine, DUNLSF, that can be obtained from IMSL. The bounding rules for this program are given in the Appendix. Fitting the growth data in Table 5.3 to the Haldane equation using the computer program in the Appendix (EPS value of 0.00001) produces values of 0.272 hr^{-1}, 121 mg/L COD, and 197 mg/L COD for μ_{max}, K_s, and K_i, respectively with an SSQ value of 0.0007. A comparison of the Haldane and Monod fits to the growth data listed in Table 5.3 is depicted in Figure 5.9. This figure and the supporting statistical data indicate that both functions provide a reasonable description of the growth data. Without additional growth rate data (at higher COD values), some engineering judgment must be applied to reconcile this somewhat equivocal situation. Considering the implications for the prediction of effluent quality, the conservative approach is to utilize the inhibitory version of the predictive model to make predictions of effluent quality. For analytical completeness, one can use both sets of constants in the predictive model and compare and contrast the

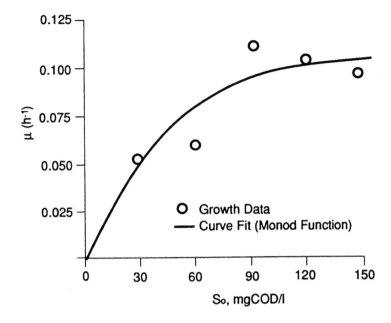

Figure 5.8. Plot comparing the experimental growth data listed on Figure 5.5 to the growth curve generated from applying the constants obtained from a computer curve fit analysis of the Monod function to Equation 1.11.

performance envelopes predicted by both the inhibitory and noninhibitory versions.

DETERMINATION OF TRUE CELL YIELD, Y_t, AND DECAY RATE, k_d

The complication with determining the values of true cell yield and decay rate is that these biokinetic constants can only be evaluated using data collected in relatively long-term pilot plant runs. Fortunately, the variation of the values of these constants has a relatively small impact on the prediction of effluent quality. The basic approach is to operate a biological treatment system at several (at least three) different growth rates (or Θ_c's). The amount of excess sludge produced (or wasted) is

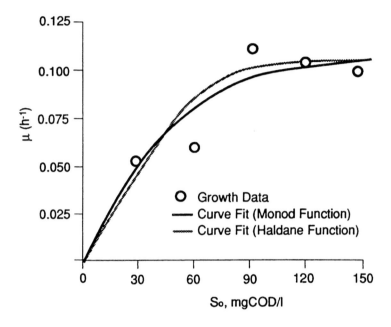

Figure 5.9. Plot comparing the experimental growth data to both the Monod and Haldane expressions using numerical values of the biokinetic constants obtained from computer curve fits of the experimental data listed on Figure 5.5.

noted at each different growth rate or Θ_c. Once sufficient data have been collected, a "maintenance plot" is made of the data in order to determine the values of Y_t and k_d. This analysis consists of a plot of observed yield, Y_o, versus net growth rate, μ_n or Θ_c. Because these analyses are relatively routine, we will not go into detail in this text but instead provide several references (Gaudy and Gaudy, 1988; Metcalf and Eddy, Inc., 1979; Rozich, Gaudy, and D'Adamo, 1983). The information in the references is sufficient to enable the reader to collect the type of data required and to the perform the ancillary analyses to compute true cell yield and decay rate. Rozich, Gaudy, and D'Adamo (1983) provide a relatively straight-forward example for determining these constants from pilot plant data.

KEY CONCEPT SUMMARY

The key concepts contained in this chapter are:

- The influent concentration of waste or waste strength, S_i, is defined as that fraction of the feed waste that can be biodegraded by the reactor biomass. This fact is embodied in the ΔCOD test and concept. The use of COD or TOC measurements to estimate S_i results in a conservative evaluation of S_i, since influent COD or TOC will always be greater than ΔCOD or ΔTOC.
- The biokinetic growth constants μ_{max}, K_s, and K_i (if the waste is inhibitory) are determined by fitting growth data (μ vs. S) to either the Monod or Haldane equation (noninhibitory or inhibitory, respectively). These growth data are generated using respirometry. The respirometric data must be "translated" into equivalent biomass concentration data before growth rates can be computed. The biokinetic growth constants have a substantial impact on the prediction of effluent quality from an activated sludge system.
- True cell yield Y_t and decay rate k_d must be determined using continuous flow tests. These constants primarily impact the prediction of waste sludge production and have a relatively minor impact on predicting effluent quality from an activated sludge system.

REFERENCES AND SUGGESTED ADDITIONAL READING

D'Adamo, P.C., Rozich, A.F., and Gaudy, A.F. Jr. (1984). "Analysis of Growth Data with Inhibitory Carbon Sources," *Biotechnology and Bioengineering, XXVI,* pp. 397–402.

Gaudy, A.F. Jr., Rozich, A.F., Moran, N.R., Garniewski, S.T., and Ekambaram, A. (1988). "Methodology for Utilizing Respirometric Data to Assess Biodegradation Kinetics." *Proceedings, 42nd Purdue Industrial Waste Conference,* Lewis Publishers, Chelsea, Michigan, pp. 573–584.

Gaudy, A.F. Jr., Ekambaram, A., and Rozich, A.F. (1989). "A Respirometric Method for Biokinetic Characterization of Toxic Wastes," *Proceedings, 43rd Purdue Industrial Waste Conference,* Lewis Publishers, Chelsea, Michigan, pp. 35–44.

Gaudy, A.F. Jr., Ekambaram, A., Rozich, A.F., and Colvin, R.J. (1990). "Comparison of Respirometric Methods for Determination of Biokinetic Constants for Toxic and Nontoxic Wastes," *Proceedings, 44th Purdue Industrial Waste Conference,* Lewis Publishers, Chelsea, Michigan, pp. 393–403.

Gaudy, A.F. Jr., and Gaudy, E.T. (1988). *Elements of Bioenvironmental Engineering,* Engineering Press, Inc., San Jose, California.

Metcalf and Eddy, Inc. (1979). *Wastewater Engineering: Treatment, Disposal, Reuse,* McGraw-Hill, New York.

Rozich, A.F., Gaudy, A.F., Jr., and D'Adamo, P.C. (1985). "Selection of Growth Rate Model for Activated Sludges Treating Phenol," *Water Research, 19,* pp. 481–490.

Rozich and Gaudy, Inc. (1988). "Manual-Instructions for Obtaining Biokinetic Data and Determining Operational Guidelines for Activated Sludge Systems," Engineering Report to the City of Baltimore, MD, Wastewater Facilities Division.

6 FACTORS AFFECTING THE VALUES OF THE BIOKINETIC CONSTANTS

INTRODUCTION

The purpose of this chapter is to review how various environmental conditions can and do impact the values of the biokinetic constants which quantify the growth potential which a biomass is capable of on a target waste. It needs to be emphasized that the system ecology and consequently the values of the biokinetic constants vary with changes in environmental conditions, waste characteristics, etc. This should not be viewed as a weakness with biological treatment technology. Conversely, with the advent of respirometric techniques that provide a cost-effective means of biokinetic constant measurement, engineers have methodology which expedites model calibration. That is, as the ecology changes, it is practicable to recalibrate the model and quantify the impact of new environmental conditions on process performance. Engineers can effectively view the values of the biokinetic constants as a measure of the "acclimation state" of the biomass. Generally speaking, the higher the maximum growth rate a population achieves on a target waste, the more well-acclimated it is to that waste. It is also useful to think of the biokinetic constants as a catalytic characterization of the system ecology. That is, it is not practicable to isolate and characterize the plethora of organisms that comprise a given biomass. However, as environmental control professionals, we are more concerned with the degradative capability that can be expected by biomass on a given waste. The quantification of the degradative capability as provided by the biokinetic constants is a kinetic means of characterizing system ecology.

The environmental conditions to be reviewed in this chapter include:

- Reactor growth rate: The rate at which a biomass is grown has a significant impact on the values of the biokinetic constants.
- Waste composition: The composition of wastewater has a huge effect on the ability of populations to degrade target components.
- Toxics: The toxic nature of a waste stream or other conditions can adversely affect the ability of a biomass to degrade wastes.
- Temperature: Temperature affects both the values of the constants and the types of wastes that can be treated.
- Population diversity: Microbial population diversity affects its ability to respond to different waste treatment situations.

We will also discuss two topics in biological treatment that are often of significance: shock loads and bioaugmentation

The topic of shock loads is important because many practitioners tend to dismiss the applicability of process models for plant design and operation because the models are derived assuming steady-state conditions while plants are generally in a dynamic situation. This section will point out ways that the process models can be employed to analyze potential shock load situations and develop compensatory strategies for dealing with these occurrences.

The topic of bioaugmentation is one which will generally solicit a jaundiced response from all but the most open-minded of environmental engineers. We will discuss the principles of bioaugmentation, how these relate to fundamentals of microbial ecology, and potential applications. Bioaugmentation in theory is a valid technological option. For example, a familiar application is the use of seed in the BOD test. Bioaugmentation has received a poor reputation in large part due to questionable practices of some companies that sell microbial products to wastewater treatment plants. It is the goal of this section to point out potential approaches that can be utilized to make this practice more akin to engineering practice than to commercial art.

SPECIFIC GROWTH RATE OF THE REACTOR

As discussed previously, specific growth rate in biological reactors is equivalent to control parameters such as F:M, Θ_c, or U. It has been observed over the years that system growth rate can significantly affect the type of biomass and consequently the ability of a reactor to handle certain wastes at certain rates. A simple example that comes to mind is nitrification. It is commonly accepted that a mean cell residence time of 8–10 d is needed to achieve nitrification. In terms of the system biomass, this means that the growth rates have to be slow enough to enable nitrifying bacteria to persist in the system at a level large enough to convert the bulk of the ammonia or total Kjeldahl nitrogen (TKN) to nitrate.

A simple guideline to remember is that the faster or slower one grows a microbial population, the higher or lower, respectively, the resulting maximum growth rate will usually be. This consequently means that the values of the biokinetic constants that influence maximum growth rate will also be affected.

As an illustration of this point, consider Figure 6.1. This figure depicts the actual and predicted values for effluent quality for heterogeneous populations growing on glucose in a chemostat. During the course of this work, several steady-state runs were made at different flow rates or

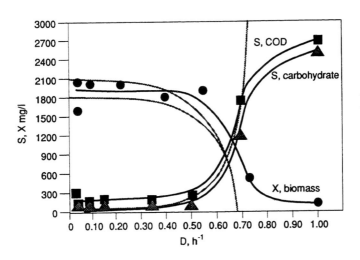

Figure 6.1. Dilute-out curves in S and X for increasing values of D (reciprocal of t̄) for heterogeneous microbial populations growing on glucose. S_i = 3000 mg/L. Dashed lines show dilute-out pattern calculated from Eqs. 2.9 and 2.10 using experimentally determined values of μ_{max}, K_s, and Y_t (k_d assumed to be 0). (Gaudy and Gaudy, 1988).

dilution rates. That is, the growth rate in the chemostat was adjusted by changing the flow rate or dilution rate. The flow rate was held constant until a steady-state condition (little variation in the observed values of reactor biomass and substrate concentration) was achieved. An example of steady-state data from this work is depicted in Figure 6.2. During each steady state, biomass was harvested from the reactor and used as inocula in separate batch growth tests. As illustrated in Figure 6.1, the predicted and observed values for effluent substrate concentration deviate from one another at higher dilution or reactor growth rates. This deviation also been observed for pure culture studies and has been attributed to the selection of subspecies which have higher μ_{max} values.

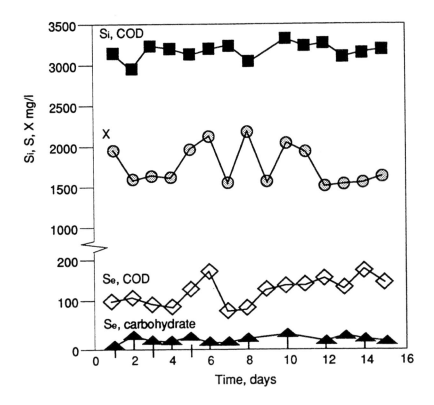

Figure 6.2. "Steady-state" values of **X** and **S**$_e$ for feed values **S**$_i$ shown in a once-through reactor operating at **D** = **0.33** h^{-1}. (Gaudy and Gaudy, 1988).

For the work described in Figure 6.1, evidence suggested that the faster reactor growth rates selected organisms which had higher μ_{max} values. The corollary argument is that the faster system growth rates washed out slower-growing species. This effect is analogous to activated sludge nitrification in that lower Θ_c values (faster reactor growth rates) wash out nitrifying bacteria, since these organisms cannot maintain sufficient growth rates to persist in the system.

A more dramatic example of the impact of reactor growth rate on the values of the biokinetic constants is found in some of the work which our group performed regarding the biodegradation kinetics of toxic or inhibitory components, specifically phenol. During these efforts, a number of different reactor types and configurations were employed in studying the biodegradation kinetics of phenol. A chemostat, an activated sludge process, and a two-stage continuous culture system were all used. This variety of reactor types enabled us to utilize a wide range of reactor growth rates for growing heterogeneous populations on phenol.

A comparison of the growth kinetics of cells utilizing phenol as a sole carbon source in a chemostat and in an activated sludge reactor is presented in Figure 6.3. The activated sludge system was operated at various steady states over a net growth rate range of 0.005 to 0.032 hr^{-1} (Θ_c range of 1.3 to 7.7 d) while the chemostat was run over a growth rate range of 0.014 to 0.054 hr^{-1} (Θ_c range of 0.80 to 3.0 d). Figure 6.3 shows that, as one would expect, the cells grown in the chemostat were capable of achieving higher growth rates, which was reflected in higher values of the maximum growth rate (μ^* for inhibitory systems). That is, as the growth rate of the reactor was increased, it increased the value of the maximum growth rate, reflected in changes of the biokinetic constants.

A more profound impact of reactor growth rate on the values of the biokinetic constants was realized by growing heterogeneous populations on phenol using the two-stage continuous culture system depicted in Figure 6.4 (Colvin and Rozich, 1986). The initial intent for using this reactor system was to develop a continuous flow technique to quantify the growth kinetics of inhibitory wastes. Since a chemostat treating an inhibitory waste will wash out once the flow rate produces a growth rate equal to the μ^* of the biomass, it is not feasible to collect continuous flow data on the right side of the inhibition curve using a chemostat alone. In a two-stage continuous culture system (shown in Figure 6.4), the flow of biomass from the first reactor enables biomass to persist in the second reactor and avoid a washout. This enables one to collect steady-state growth data on an inhibitory substrate at relatively high steady-state

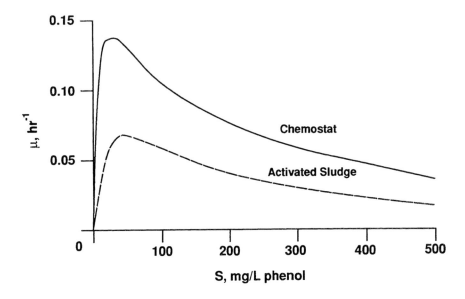

Figure 6.3.   Comparison of phenol growth kinetics from biomass grown
in a chemostat (μ_{max} = 0.19 hr^{-1}, K_s = 7.9 mg/L, and K_i =
139 mg/L, Colvin and Rozich, 1986) and in an activated
sludge system (μ_{max} = 0.194 hr^{-1}, K_s = 48 mg/L, and K_i =
62 mg/L, Rozich and Gaudy, 1985).

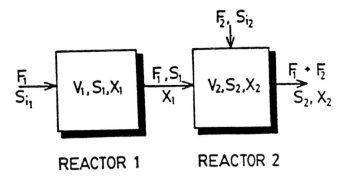

Figure 6.4.   Schematic of two-stage continuous culture system. (Colvin
and Rozich, 1986)

substrate concentrations which is not possible in a chemostat because of washout due to exceeding the μ^* criterion.

Heterogeneous populations grown on phenol in the second stage of the two-stage continuous culture system responded with greatly enhanced growth rates, as illustrated in Figure 6.5. This figure shows that the maximum growth rate (μ^*) of the cells grown in the second stage of this system was approximately four times higher than that of the rate exhibited by the organisms in the first stage. The technique of forcing higher growth rates in the second stage of this system selected a biomass with greatly enhanced growth kinetics, evidenced by the contrast in values of the biokinetic constants.

For additional comparative purposes, consider Figure 6.6. This figure simply plots the data in Figures 6.3 and 6.5 on one graph. The purpose of this figure is to illustrate the range of potential growth rates and values of the biokinetic constants that are possible for heterogeneous populations metabolizing phenol. The changes in the values of the biokinetic constants were achieved by selecting different microbial populations using a variety of reactor configurations to manipulate system growth rate.

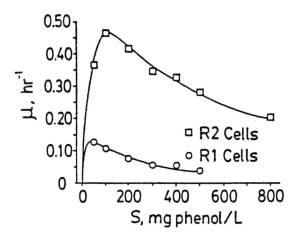

Figure 6.5. Plot of average growth rates detected in batch growth studies fitted to the Haldane Equation. Resulting values for μ_{max}, K_s, and K_i were 0.19 hrs^{-1}, 7.9 mg/L and 1.39 mg/L, respectively, for the R1 Cells and 1.07 hr^{-1}, 79 mg/L, and 172 mg/L, respectively, for the R2 Cells. (Colvin and Rozich, 1986).

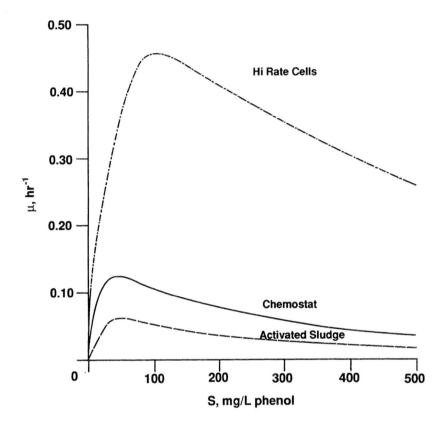

Figure 6.6. Comparison of kinetics of cells grown on phenol in a chemostat (Figure 6.3) and in activated sludge (Figure 6.3) with kinetics of cells grown in a high-rate system (μ_{max} = 1.07 hr^{-1}, K_s = 79 mg/L, and K_i = 172 mg/L, Colvin and Rozich, 1986).

As a side note, it should be emphasized that the high growth rates were achieved on the inhibitory or "toxic" organic phenol by simply using heterogeneous populations and a relatively novel but readily usable culturing technique. The point is that developing biological wastewater systems to degrade high-strength organic waste streams rapidly does not require the use of "mutant" bacteria or genetically engineered organisms. By employing the principles of reactor engineering and the knowledge that the range of growth kinetics of natural microbial systems is usually larger than we suspect, environmental engineers and scientists can devise biological waste treatment systems that are capable of a much greater performance envelope than what is currently thought to be achievable.

WASTE COMPOSITION

Waste composition is a factor that one would intuitively expect to exert an effect on the values of the biokinetic constants and consequently on the performance of an activated sludge system. Specifically, this section refers more to variations in biodegradable carbon source than to the presence of materials, such as heavy metals, that may adversely affect the ability of the biomass to degrade wastes; the impact of substances such as heavy metals on the values of the biokinetic constants will be discussed in the "Toxics" section. If one is concerned about the potential impact of waste composition changes on the performance of a biological treatment system, it is a simple task to evaluate the potential changes in the values of the biokinetic constants and to use the model to predict the impacts on plant performance.

An important point about a target wastewater is that the effects of the composition are not readily predictable. For example, an old axiom in the environmental field was that a biodegradable inhibitory waste stream such as phenol was easier to handle with the presence of a more degradable carbon source such as glucose. From a reactor engineering perspective, this implies that the presence of the alternate carbon source gives the biomass a stronger capacity to metabolize the phenol; this particular case lends itself as a good illustrative example regarding the difficulty of predicting the impact of waste composition on biomass capability to degrade target components.

Our group became interested in the situation presented in the preceding paragraph. That is, is it appropriate and/or beneficial to mix an easily biodegradable carbon source with a difficult or inhibitory one in order to enhance the treatability of the inhibitory carbon source? This work (Rozich and Colvin, 1986) was a corollary to other efforts conducted regarding the delineation of the biodegradation kinetics of phenol and other toxic or inhibitory compounds.

For this work, two different types of biomass were cultivated in bench-scale reactors: one acclimated to phenol as a sole carbon source and one acclimated to a mixed carbon source consisting of phenol and a more easily degradable compound, glucose. An initial set of tests were conducted to assess the influence of glucose on the ability of the cells solely acclimated to phenol to degrade phenol. Representative results are shown in Figures 6.7 and 6.8. In each of these tests, two concurrent batch reactors were run using cells acclimated only to phenol; one reactor received only phenol while the other reactor received a mixed waste consisting of phenol and glucose. Interestingly enough, these results show that the glucose, which is representative of an easily degradable waste, hinders phenol removal. This is stated because the results showed that phenol removal rates slowed down in the reactors that received the glucose. Additionally, and somewhat surprising, the data show that the cells acclimated only to phenol preferentially removed phenol while not metabolizing the glucose. These data, along with other work in our laboratories, suggested that adding a more easily biodegradable carbon source to a toxic waste can actually decrease the ability of the biomass to degrade the target waste. These statements apply, of course, to a biomass that degrades a toxic organic waste (e.g., phenols and other aromatics) when using it as a sole source of carbon and energy and not to co-metabolic situations.

Other tests were performed using the biomass which was acclimated to a mixture of phenol and glucose. Typical results are shown in Figures 6.9 and 6.10. These tests showed, as one would expect, that the biomass preferentially removed glucose. That is, with the shift in acclimating

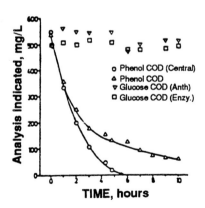

Figure 6.7. Mixed phenol/glucose substrate removal by phenol acclimated cells. Initial biomass concentration = 150 mg/L. Glucose assessed utilizing both anthrone and enzymatic methods. (Rozich and Colvin, 1986).

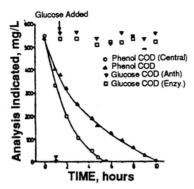

Figure 6.8. Mixed phenol/glucose substrate removal by phenol acclimated cells. Glucose fed to mixed waste reactor 1.5 h after cells were actively utilizing phenol. Initial biomass concentration = 150 mg/L. Glucose assessed utilizing both anthrone and enzymatic methods. (Rozich and Colvin, 1986).

conditions, a change was realized in the acclimation state of the biomass that resulted in the preferential use of the more easily degradable waste. However, these results and others did not show that the biomass exhib-

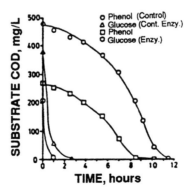

Figure 6.9. Mixed phenol/glucose substrate removal by phenol/glucose acclimated cells subjected to nonproliferating conditions. Initial biomass concentration = 400 mg/L. (Rozich and Colvin, 1986).

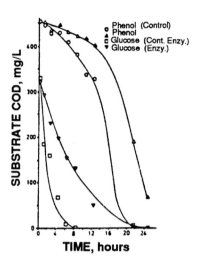

TIME, hours

Figure 6.10. Mixed phenol/glucose substrate removal by phenol/glucose acclimated cells. Initial biomass concentration = 300 mg/L. (Rozich and Colvin, 1986).

ited an increase in the values of the biokinetic constants on phenol. Conversely, this line of work in our laboratory indicated that populations acclimated to both phenol and glucose had a decreased ability to degrade phenol as indicated by the values of the biokinetic constants.

The above statements are reinforced by the comparative kinetic results, which are shown in Figure 6.11. This figure compares the kinetic characteristics of phenol on organisms that were cultivated in a two-stage continuous culture system; the lower kinetic response was demonstrated by the cells accclimated to a mixture of phenol and glucose while the greater growth potential on phenol was realized by the biomass grown using phenol as a sole source of carbon. The bottom line is that the addition of a more easily degradable carbon source is not necessarily going to make a toxic waste more degradable, unless, of course, there is a co-metabolic requirement for the alternate carbon source to achieve biodegradation of the target toxic organic. In point of fact, the addition of a more easily degradable carbon source may actually lower the kinetic ability, as quantified by the values of the biokinetic constants, of a biomass to degrade certain difficult-to-degrade or toxic wastes.

Another reason why waste composition effects are not readily predictable is due to the potential effect of metabolic control mechanisms

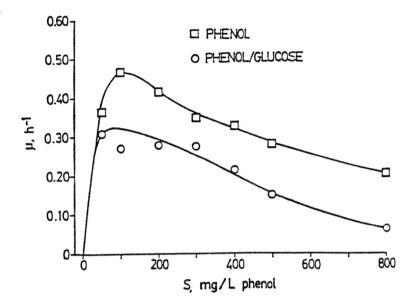

Figure 6.11. Comparison of phenol batch growth rates for cells cultured using phenol as a sole carbon source and using the mixed phenol/glucose waste. (Rozich and Colvin, unpublished data).

(Gaudy and Gaudy, 1980). Metabolic control mechanisms occur when a microbial system will preferentially degrade one carbon source over another. In one aspect of these mechanisms, organisms repress the manufacturing of enzymes needed for degrading a target carbon source because of the presence of an alternate source. This is termed repression because the cells attenuate the synthesis of enzyme required for metabolizing the one carbon source. In another aspect, the presence of an alternate carbon source results in a lowering of enzyme activity. This is termed catabolite inhibition when the alternate carbon source inhibits the activity of the enzymes needed for degrading the target carbon source (Gaudy and Gaudy, 1980).

The effects of metabolic control mechanisms in batch systems are illustrated in Figure 6.12. Cells were subjected to resting conditions wherein the nitrogen source was withheld to prevent net synthesis of protein. This figure shows that rapid blockage of the removal of both

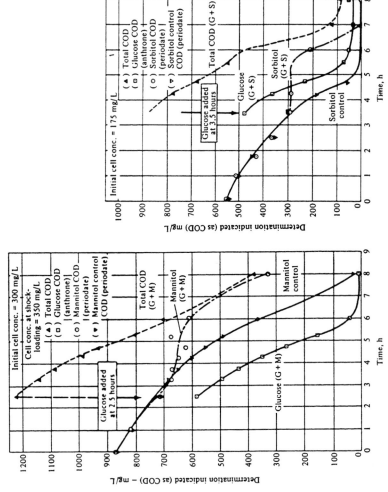

Figure 6.12.   Blockage of removal of mannitol (left) and sorbitol (right) by glucose, which was injected as a slug dose into the medium during growth on the sugar alcohols. (Gaudy, Komolrit, and Gaudy, 1964).

mannitol and sorbitol was achieved after injection of glucose into the medium, which demonstrates the attenuation of enzyme function due to the presence of glucose. Mannitol and sorbitol utilization did not resume until the glucose was metabolized.

The impact of metabolic control mechanisms may be most relevant regarding qualitative shock loads (i.e., situations involving a sudden, radical change in the composition of the influent wastewater). As an example, consider Figure 6.13. This figure depicts the result of changing the influent waste stream from glycerol to a combination of glycerol and glucose and the impact of changing a waste stream from sorbitol to a mixture of sorbitol and glucose. The results show that in addition to leaking glycerol and sorbitol in the respective systems, both reactors also leaked a significant quantity of product (nonsubstrate) COD. These results suggest that a sudden and significant change in waste quality will cause leakage of both influent waste COD and product COD.

TOXICS

Many times the presence of substances such as heavy metals, non-biodegradable organics, salts, etc. can by themselves or in conjunction with other materials produce a toxic effect on the ability of a biomass to degrade target wastes. This effect differs from the previously discussed phenomenon of substrate inhibition in that the toxic material in this case is not being biodegraded, but it exerts a detrimental effect on the ability of the biomass to metabolize the target waste. As one should expect, this toxic effect, which presumably will depress system performance, is predictable if the impact of the toxic materials on the values of the biokinetic constants is quantified.

As a generic example, consider Figure 6.14. This figure depicts two idealized curves showing biomass growth kinetics, i.e., μ as a function of waste or substrate concentration. The curve showing the higher growth rates represents the kinetic ability of the biomass without the presence of the toxic material; this represents the baseline case. The other curve quantifies the degradative capability of the biomass in the presence of a toxic material and shows, as would be expected, lower growth rate potential. Once these data are generated, the values of the biokinetic constants are inserted into the process model and the impact of the toxic material on the performance or the operating envelope of the biological waste treatment system is quantified.

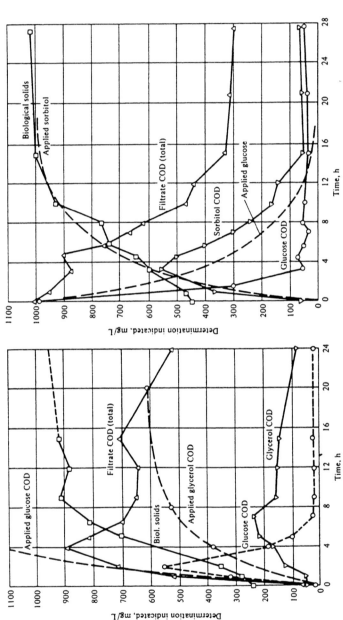

Figure 6.13. Shock load response of heterogeneous microbial populations growing under steady state conditions at D = $^1/_4$ h^{-1} prior to a change in inflowing substrate. (Left) At 0 h, the feed composition was changed from 500 mg/L of glycerol to 500 mg/L of glycerol plus 1500 mg/L of glucose. The dashed line, applied glucose COD, represents the glucose COD that would have been present if it were not metabolized. Applied glycerol COD is the concentration of glycerol COD that would have been present if the metabolism of glycerol had stopped at the time of administering the shock. (Right) At 0 h, the feed composition was changed from 1000 mg/L of sorbitol to 2000 mg/L of sorbitol, and 1000 mg/L glucose was injected directly into the reactor as a slug dose. The dashed lines labeled applied glucose and applied sorbitol represent the concentrations of the two compounds that would have been present if they had not been metabolized. (Komolrit and Gaudy, 1966).

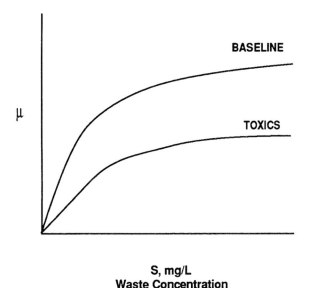

μ

BASELINE

TOXICS

S, mg/L
Waste Concentration

Figure 6.14.   Idealized growth curves illustrating the impact of toxicants on growth kinetics.

As specific examples, we will show the results of the impact of toxic effects for three activated sludge systems:

- A domestic waste treatment system that was affected by an aggregate impact of several industrial discharges.
- An industrial waste treatment facility that was handling an easily biodegradable organic waste but that had to contend with the effects of the heavy metal cobalt.
- An oxygen-activated sludge facility that was plagued by carbon dioxide inhibition due to CO_2 accumulation in the reactor headspace.

Figure 6.15 compares the kinetics of the biomass in a municipal waste treatment plant receiving a relatively heavy industrial waste input with that of a system with a light industrial contribution. For some time, the staff of the former plant had been complaining that the plant was being besieged by toxic inputs. However, when kinetic analyses were performed, they indicated that the growth kinetics adhered to the "conven-

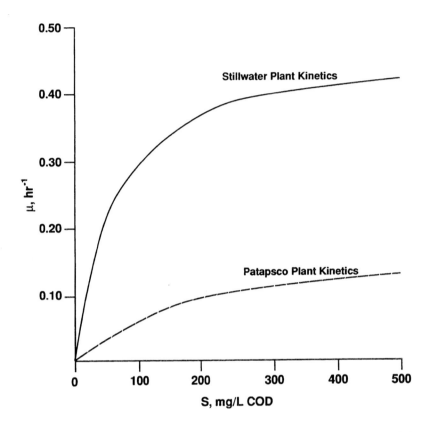

Figure 6.15. **Comparison of kinetics for a municipal plant receiving little industrial waste (μ_{max} = 0.50 hr^{-1}, K$_s$ = 63 mg/L) (Stillwater) and one receiving a heavy industrial waste contribution (Patapsco) (μ_{max} = 0.18 hr^{-1}, K$_s$ = 176 mg/L).**

tional" or Monod-type kinetics. This made it somewhat troublesome to argue that the plant was in a difficult treatment situation that was not representative of a "typical" domestic waste. The comparison in Figure 6.15 of the values of the biokinetic constants of the former plant with those of the plant receiving a light industrial input quantified the difference in biodegradative characteristics, enabling us to make a convincing case that this treatment situation was not comparable to a typical domestic waste and that the design and operation of the plant warranted special consideration. Subsequent process analyses using the model and the values of the biokinetic constants that were generated as part of this work reinforced this point and defined system design and operating requirements (Rozich and Gaudy, Inc., 1986).

The case of the cobalt was more straightforward. For this work (Constable et al., 1991), a biomass was acclimated to a synthetic waste mixture that simulated the organic composition of the influent waste stream. Two comparative biokinetic tests were performed using the acclimated biomass; one test utilized the organic waste mixture and the other test employed the organic waste mix, which was amended with 1 mg/L of cobalt. The results are depicted in Figure 6.16. This figure shows that the presence of the cobalt cut the biodegradative potential of the biomass by almost a factor of two. As was the case with the previous example, follow-on process analyses using the values of the biokinetic constants and the process model quantified the impact of the presence of the cobalt on both plant performance and capacity.

A case of CO_2 inhibition presents an interesting example of a toxic effect inhibiting biomass ability to degrade waste. The efficient transfer of oxygen in oxygen-activated sludge systems is due to the fact the oxygen is delivered as a relatively pure stream whereas in conventional sys-

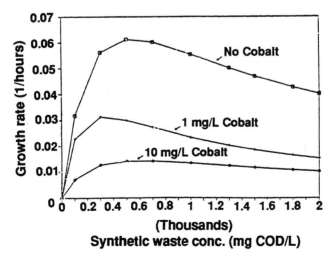

Figure 6.16. A series of respirometer studies were performed to quantify the impact of cobalt on the operation of a biotreatment process (Rozich and Colvin, 1990).

tems air is the carrier gas. For process economy reasons (generation of pure oxygen is expensive), activated sludge reactors in these systems are covered to maintain a relatively high oxygen concentration in the carrier gas in order to maintain efficient oxygen transfer kinetics. Covering the reactors also means that CO_2 concentrations can accumulate in the headspace above the liquid level of the reactors, which results in relatively high CO_2 concentrations in the mixed liquor. Work performed at the Patapsco Wastewater Treatment Plant in Baltimore, Maryland (Martin, 1988) demonstrated the adverse effects of CO_2 inhibiton on the kinetic capability of the biomass to remove COD. Figure 6.17 shows the results of tests performed to measure the influence of the percent of carbon dioxide saturation on inhibiting COD removal by the plant biomass. Although the data were somewhat scattered, they did indicate a clear impact of CO_2 on COD removal. Other ancillary tests were performed to validate that the inhibition was due in large part to the elevated levels of CO_2 in the mixed liquor and not attributable to depressed pH, which will also result from CO_2 accumulation.

Various toxic materials can by themselves or in synergy with other components depress the ability of system biomass to degrade target wastes. This effect can be quantified via comparative evaluation of the biokinetic constants. This information can then be used in conjunction with the process model to provide quantitative data for making management decisions regarding the removal or pretreatment of the toxic materials and potential cost/benefits for the purpose of enhancing the performance of the biological waste treatment system.

TEMPERATURE

Temperature is well-recognized by microbial kineticists as having a significant effect on microbial growth kinetics. Nitrification is a good example of temperature effects. Nitrifying organisms have a relatively small temperature range with 10°C being the low end and approximately 45°C defining the high end. Consequently, operators are concerned when they must achieve winter nitrification. In contrast, one of the advantages of the autothermal aerobic digestion (ATAD) process is that the high operating temperatures suppress nitrification, which means that these systems have a lower aeration demand than conventional aerobic digestion.

Practically speaking, the basic approach for quantifying the impacts of temperature on the treatment capability of a system is to delineate the

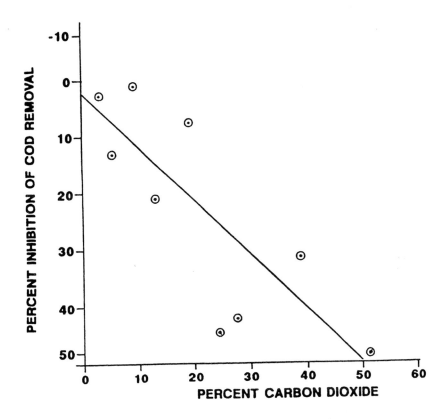

Figure 6.17. Inhibitory effect of increasing carbon dioxide concentration on removal of chemical oxygen demand by activated sludge (Martin, 1988).

effect of temperature on the values of the biokinetic constants. A good case in point is some work which was performed on the biodegradation of a landfill leachate (Rozich and Colvin, 1988). The engineers on the project were concerned with the potential performance of the system under winter conditions. The landfill was located in Canada where low temperatures (~ 10°C) are commonplace. The methodology employed consisted of evaluating the biokinetic constants at normal temperatures of 20°C and comparing these results with a kinetic evaluation performed at 10°C. (Most respirometers have the feature of temperature control and obtaining kinetics at different temperatures is relatively easy.) The results of the comparative kinetic evaluations are shown in Figure 6.18 and indicate, as one would expect, that the lower temperatures exert a significant impact on the values of the biokinetic constants; these data are then used to predict the effect of lower temperatures on system performance.

Finally, some comments are in order regarding the application of autothermal systems for high-strength ($\geq 30,000$ mg/L COD) waste treatment. Autothermal aerobic digestion systems (or ATAD) are self-heating aerobic biological treatment processes which operate at high temperature ranges (45°C to 70°C). Needless to say, microbial degradation rates at these temperatures will be quite high, and higher temperatures will also result in lower sludge production because cell yields drop with increased temperatures.

Conventional engineering strategy tends to lend itself toward utilizing

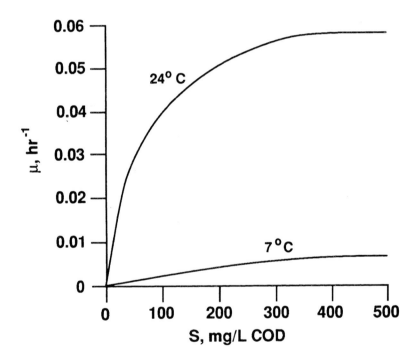

Figure 6.18. Impact of temperature on kinetic response of activated sludge acclimated to landfill leachate.

anaerobic systems for high-strength waste applications. If methane utilization is economical, then this seems to be the prudent choice. However, it should be pointed out that autothermal aerobic systems can be modeled with the same approach as that used for aerobic systems, which operate at conventional temperatures. (That is, operate a bench-scale pilot reactor, evaluate the biokinetic constants utilizing respirometry, and use the model to perform the value engineering analyses of the system.)

We suggest that environmental technologists consider the autothermal aerobic process as a potentially viable alternative because work in our own laboratories (Gaudy, Rozich, and Garniewski, 1987) indicated the potential for improved treatment for a high-strength landfill leachate when utilizing an autothermal thermophilic system. The key point is that the predictive tools are available for engineering high-temperature aerobic treatment processes. Other work (Rozich, A.F., Clay (nee Garniewski), S.G., and Colvin, R.J., 1991) showed that a bench-scale thermophilic aerobic system operating at 50°C demonstrated excellent COD removal when treating a high-strength ground water containing 200,000 mg/L COD. Ancillary respirometric tests were performed at 50°C to quantify kinetics and define the process operating envelope of the system. Thus, ATAD systems may warrant more serious consideration for treatment applications that have traditionally been exclusively reserved for anaerobic technology.

POPULATION DIVERSITY

Population diversity refers to the diversity of the ecology that makes up a biomass. A pure culture consisting of one species of micoorganism is essentially devoid of diversity. In contrast, an activated sludge system with a long Θ_c will not only have a plethora of single-cell bacteria, fungi, protozoa, etc., but it will also have multicelled creatures, which are representative of the highest trophic level in the system ecology. The level of diversity exerts an influence on the types of waste a biological system can handle and the rates at which the wastes can be processed; this, of course, is quantified via the biokinetic constants. In general, it is probably safe to state that the more specialized, or less diverse, a biomass is, the greater the growth kinetics will be on a target waste or compound; that is, the biomass will have an enhanced capability to degrade a particular waste stream or compound. As a corollary, it is also probably safe to state that as populations grow more diverse, they gain an enhanced capability to deal with a greater diversity of target compounds in a waste stream. Examples will be discussed below.

In a previous section of this chapter, we discussed the use of a two-stage continuous culture system depicted in Figure 6.4 to study the kinetics of heterogeneous populations metabolizing phenol. It is interesting to note that workers (Jones, 1973) used this same reactor system to evaluate the inhibition kinetics of phenol of a pure culture of bacteria. This group did not report enhanced phenol degradation rates in the second stage of the two-stage reactor. That is, the cells growing in the second stage of the system shown in Figure 6.4 at high phenol concentrations demonstrated inhibited growth kinetics and exhibited lower growth rates than the organisms in the first stage of the system. This contrasts sharply to the results observed for heterogeneous populations which were cultivated on phenol in the two-stage system; that is, the heterogeneous, or more diverse, biomass exhibited substantially enhanced growth kinetics in the second stage of the reactor system depicted in Figure 6.4.

Figure 6.19 compares the phenol growth kinetics of the pure culture and those of the heterogeneous cells cultivated in the second stage of the two-stage system. This figure quantifies the differences in growth kinetics between the pure culture and the heterogeneous biomass. It is evident that using a more diverse heterogeneous population for the two-stage continuous culture work produced a biomass with enhanced growth kinetics on phenol, due in large part to the fact that the use of a more diverse culture provided a larger pool of microbial species with which to select cells that exhibited enhanced phenol degradation kinetics. As previously stated, less diverse populations are less likely to be capable of handling relatively diverse waste streams. Consider the data presented in Figures 6.7 and 6.8; these figures show the response of a biomass that was acclimated to phenol as a sole source of carbon to a mixed waste consisting of phenol and glucose. This biomass can be considered as relatively specialized and lacking in diversity, especially when compared to the populations grown in, for example, a domestic wastewater treatment facility which has a waste stream with a large variety of carbon sources. These figures show that, although the added carbon source, glucose, is easily degradable, the biomass ignored and preferentially metabolized the phenol. If this were a full-scale plant, the glucose would show up in the plant effluent as BOD and the facility would likely experience a permit violation. Thus, the more specialized phenol populations were not capable of handling a more diverse waste stream consisting of phenol and the relatively easily degradable compound, glucose.

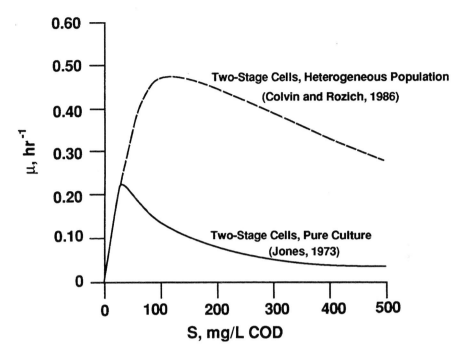

Figure 6.19. Comparison of phenol growth kinetics in a two-stage system of heterogeneous populations (μ_{max} = 1.07 hr^{-1}, K$_S$ = 79 mg/L and K$_i$ = 172 mg/L, Colvin and Rozich, 1986) and a pure culture (μ_{max} = 0.29 hr^{-1}, K$_s$ = 1 mg/L, and K$_i$ = 110 mg/L, Jones, 1973).

SHOCK LOADS

The topic of shock loads conjures up feelings of dread for any wastewater treatment plant manager or operator, especially those involved with biological wastewater treatment facilities. The purpose of this section is to provide some discussion regarding the key features of shock loads as they relate to biological wastewater treatment operations. We will then discuss ways in which the modeling approach described in this book can be applied to formulate proactive strategies that can attenuate process upset due to a shock loading situation.

It is first prudent to define what we mean when we say "shock load." There are generally three different types of shock load:

- qualitative
- quantitative
- toxic

A qualitative shock load consists of a radical compositional change in the nature of the target compounds in the influent waste stream. For example, the results depicted in Figures 6.7 and 6.8 are an example of a qualitative shock load to the phenol-acclimated cells since this biomass was challenged with a mixed phenol/glucose waste. A quantitative shock load entails a quantitative mass loading increase in the target waste components in the influent waste stream; this increase can be caused hydraulically by an increase in waste flow rate or by an increase in the concentration of the target waste components. A toxic shock load involves the sudden discharge of substances (e.g., heavy metals, high salt concentrations, etc.) that can depress the ability of the biomass to degrade the target wastes.

The potential response of the biological system to all of the above situations can be assessed by evaluating the impact of the new waste condition on the values of the biokinetic constants. Once the constants are defined, they are inserted along with appropriate loading and operational information into the process model to predict the system capacity or performance envelope that can be expected for the new conditions.

A common criticism of model application is that the model presented herein is derived assuming steady-state conditions, while a shock load situation is a dynamic event requiring a model to account for the dynamic conditions. This criticism is valid in part, because the model is in fact derived using a steady-state assumption, but this does not invalidate the application of the model for predicting ultimate system response to a new operating condition or for formulating strategies vis-á-vis modifications of operational parameters to buttress the plant against potential upset due to a shock load event. The key point to recognize is that a shock load event entails going from one steady state to another; the pre-shock condition represents the "old" steady state while the postshock condition represents the new steady state. The new set of biokinetic constants defines a "worst case" scenario for the biomass at the new condition since the biomass has not had a chance to acclimate to the waste change. Using the biokinetic constants, the model, and the new loading condition, we can predict system response performance at the new steady state.

It should be noted that a substantial amount of investigational work (Gaudy and Gaudy, 1980) has shown that a biomass frequently demonstrates an extra assimilative capacity during a shock load situation. That is, extensive controlled shock load tests have shown that systems do not immediately leak soluble substrate during the transient state between the old and new steady-state conditions. An example is given in Figure 6.20. This figure shows a bench-scale activated sludge reactor that was treating a waste stream consisting of glucose as the sole carbon source and subjected to a sixfold quantitative shock load in influent COD concentration (S_i went from 500 to 3000 mg/L). It is evident in this figure that although there was effluent deterioration due to suspended solids in the clarifier effluent (an effect which we are not able to predict), little leakage of COD occurred; the data suggested that the extra capacity was due to an oxidative assimilation mechanism that enabled the cells to take in and store the excess carbon during the shock load condition. This mechanism attenuated soluble COD leakage and provided an extra "buffer" not predicted by the model.

Conclusions regarding "extra" assimilative capacity may be questioned because the test results shown in Figure 6.20 were generated using the relatively biodegradable carbon source, glucose. Figure 6.21 shows the results for a bench-scale activated sludge process that was treating the toxic organic phenol and was subjected to a twofold quantitative shock load (S_i went from 1000 mg/L to 2000 mg/L phenol) (Rozich and Gaudy, 1985). As indicated by the data, the system exhibited little leakage of soluble COD or phenol; however, 7 d after initiation of the shock load condition, the system realized a catastrophic failure manifested by massive leakage of both soluble COD and phenol (800 mg/L). The steady-state version of the model accurately predicted this failure. We attributed the observed extra assimilation capacity, which "bought time" before the failure, to oxidative assimilation; other work in our laboratories (Rozich and Lowe, 1985) showed that microorganisms can remove phenol from solution via this mechanism.

The key point with the examples in Figures 6.20 and 6.21 is that work to date indicates that cells will frequently exhibit an extra sorptive capacity with regard to substrate or waste removal during the transient in a shock load situation. This means that model predictions for process performance in the new steady state will be conservative with respect to soluble COD leakage; that is, the model predictions can be utilized to "bound" expected performance results for the aftermath of a shock load or, from a proactive posture, be used to formulate strategies to attenuate the likelihood of permit violation due to COD leakage. It should be noted that the model presented herein or any other model is not capable of predicting the impact of a shock load to biomass settlability and the

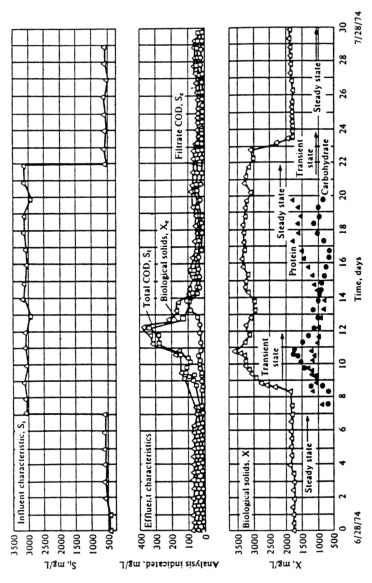

Figure 6.20. Response of an activated sludge system operating with $\alpha = 0.25$ and $X_R = 8000$ mg/L to step changes in glucose concentration, 500 to 3000 to 500 mg/L. (Adapted from Saleh and Gaudy, 1978)

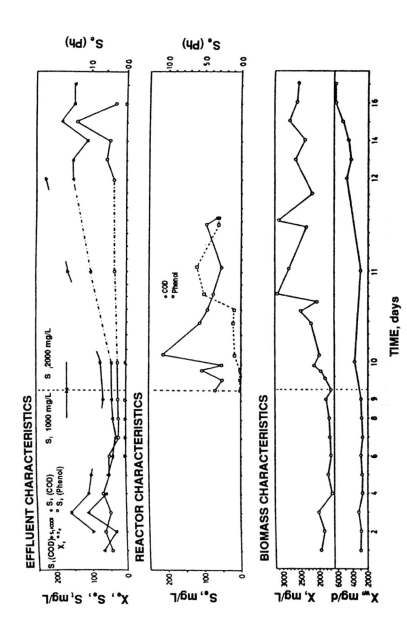

Figure 6.21. Loss of biomass and substrate from the reactor 1 week after applying a step increase in phenol concentration from 1000 to 2000 mg/L. (Rozich and Gaudy, 1985).

subsequent potential for suspended solids losses in the secondary clarifier. If settlability problems persist, it is recommended that alternate strategies (e.g., the use of chemical or flocculant aids or the use of selector technology) be evaluated to address this situation.

BIOAUGMENTATION

Bioaugmentation generally refers to the use of microbial additives for the stated goal of enhancing the performance of biological wastewater treatment systems. Operationally, it can be thought of as any situation where microbes are introduced to enhance or speed up a biological reaction. The BOD test is a good example, since seed from a primary clarifier effluent is added to a BOD bottle to insure that sufficient oxygen depletion occurs in a 5-d period.

The efficacy of bioaugmentation has been shown to be valid under controlled conditions. For example, Edgehill and Finn (1983) demonstrated that a bench-scale activated sludge system that was dosed with a small side stream of organisms preacclimated to pentachlorophenol (PCP) could take a PCP shock without leaking PCP. A duplicate unit that did not receive the organisms responded to the shock situation by leaking a substantial portion of the PCP. This laboratory demonstration clearly showed that the concept of bioaugmentation is tenable.

Assuming that the concept of bioaugmentation is valid, the next step is to determine if a product ("bugs" from a vendor) will achieve any more on a particular waste stream or component than the biomass which is indigenous to a particular reactor. This is a performance question and can be answered by determining the kinetics of both the indigenous reactor biomass and the product on the target waste or stream; the respirometric methodologies and ancillary analyses described in Chapter 5 can be used to quantify the μ_{max}, K_s, and K_i (inhibitory situation) of each biomass. Care must be exercised in performing these tests to insure that each the initial biomass concentrations are quantified. For example, if one were to test a product with an inordinately high concentration of cells, the sample would show superior kinetics based on respirometry alone. However, once the data are translated into equivalent biomass concentrations and the corresponding growth rates are calculated, a true kinetic comparison can be made.

The key point is that, in some cases, bioaugmentaton may be a useful operational tool for difficult streams or as backup to existing operational protocols. The first question is whether the product organisms are any better at degrading the target streams than the existing biomass; this question can be answered with a comparative respirometric test and ancillary kinetic analyses. The next question is economical. Is it economical to use the bioaugmentation product vs. using some other design or operational technique to achieve the target treatment goals?

KEY CONCEPT SUMMARY

The key concepts contained in this chapter are:

- The values of the biokinetic constants are all influenced by various environmental and operational conditions. The key point is that this variation can be easily quantified using the respirometric techniques described in Chapter 5. The values of the biokinetic constants and the engeineering models presented in Chapter 3 enable the process analyst to quantify the performance envelope for a biological treatment system for a target environmental or operatonal condition.
- The growth rate of a biological reactor (or the equivalent F:M or Θ_c) imparts a significant impact on the values of the biokinetic constants. Faster-growing systems will have a tendency to select for populations with higher μ_{max} or μ^* (if the waste is inhibitory) values; the opposite also holds for slower-growing systems. Although a system may be ultimately capable of attaining a higher μ_{max} or μ^* value, it is important to realize that reactor performance is dictated by kinetic capability of the organisms which are currently inhabiting the reactor.
- The composition of the waste stream plays a significant role in determining the kinetic capability of a biomass to handle certain components.
- Toxics such as heavy metals, organics, or high concentrations of salt impact reactor performance by depressing kinetic capability. Temperature also exerts a significant effect on the values of the biokinetic constants.
- Greater diversity in a microbial population gives it the capability to degrade a greater variety of waste components and to respond to changing waste treatment situations.

REFERENCES AND SUGGESTED ADDITIONAL READING

Colvin, R.J., and Rozich, A.F. (1986). "Phenol Growth Kinetics of Heterogeneous Populations in a Two-Stage Continuous Culture System," *J. Water Poll. Control Fed., 58*, pp. 326–332.

Constable, S.W., Rozich, A.F., DeHaas, R., and Colvin, R.J. (1991). "Respirometric Investigation of Activated Sludge Bioinhibition by Cobalt/Manganese Catalyst" *Presented, 46th Annual Industrial Waste Conference*, Purdue University, West Lafayette, IN, May 1991.

Edgehill, R.V., and Finn, R.K. (1983). "Isolation, Characterization, and Growth Kinetics of Bacteria Metabolizing Pentachlorophenol," *Eur. J. Appl. Microbiol. Biotechnol., 16*, pp. 179–189.

Gaudy, A.F. Jr., and Gaudy, E.T. (1980). *Microbiology for Environmental Scientists and Engineers*, McGraw-Hill, New York.

Gaudy, A.F. Jr., and Gaudy, E.T. (1988). *Elements of Bioenvironmental Engineering*. Engineering Press, Inc., San Jose, California.

Gaudy, A.F., Jr., Komolrit, K., and Gaudy, E.T. (1964). "Sequential Substrate Removal in Response to Qualitative Shock Loading of Activated Sludge Systems." *Appl. Microbiol., 12*, pp. 280–286.

Gaudy, A.F. Jr., Rozich, A.F., and Garniewski, S.T. (1987). "Biological Treatment of Concentrated Landfill Leachate," *Proceedings, 41st Annual Purdue Industrial Waste Conference*. Lewis Publishers, Inc., Chelsea, Michigan, pp. 627–638.

Jones, G.L. (1973). "Substrate Inhibition of Bacterium NCIB 8250 by Phenol," *J. Gen. Microbiol., 74*, pp. 139–149.

Komolrit, K., and Gaudy, A.F., Jr. (1966). "Biochemical Response of Continuous-Flow Activated Sludge Processes to Qualitative Shock Loadings." *J. Water Poll. Control Fed., 38*, pp. 85–101.

Martin, J.K. (1988). "Inhibition of Respiration in Activated Sludge by High Carbon Dioxide Concentration – A Laboratory Study," Public Technology, Inc., Washington, D.C.

Ramanathan, M., and Gaudy, A.F. Jr. (1969). "Effect of High Substrate Concentration and Cell Feedback on Kinetic Behavior of Heterogeneous Populations in Completely Mixed Systems," *Biotech. Bioeng., 11,* pp. 207–237.

Rozich, A.F., and Colvin, R.J. (1986). "Effects of Glucose on Phenol Biodegradation by Heterogeneous Populations," *Biotechnology and Bioengineering, XXVII*, pp. 965–971.

Rozich, A.F., and Colvin, R.J. (1988). Unpublished results.

Rozich, A.F., and Colvin, R.J. (1990). "Formulating Strategies for Activated Sludge Systems," *Water Engineering & Management, 137*, 10, pp. 39–41.

Rozich, A.F., Clay (nee Garniewski), S.G., and Colvin, R.J. (1991). Unpublished results.

Rozich and Gaudy, Inc., (1986). "Determination of the Numerical Values of the Biokinetic Constants and Implications to the Design of the Expanded Facilities for the Patapsco Wastewater Treatment Plant," Engineering Report to the City of Baltimore, MD, Wastewater Facilities Division.

Rozich, A.F., and Gaudy, A.F., Jr. (1985). Response of Phenol Acclimated Sludge to Quantitative Shock Loading. *J. Water Poll. Control Fed., 57*, pp. 795–804.

Rozich, A.F., Gaudy, A.F. Jr., and D'Adamo, P.C. (1983). Predictive Model for Treatment of Phenolic Wastes by Activated Sludge, *Water Research, 17*, pp. 1453–1466.

Rozich, A.F., Gaudy, A.F., Jr., and D'Adamo, P.C. (1985). "Selection of Growth Rate Model for Activated Sludges Treating Phenol," *Water Research, 19*, pp. 481–490.

Rozich, A.F., and Lowe, W.L. (1984). "Oxidative Assimilation Treatment of a Nitrogen-Deficient Toxic Waste," *Biotechnology and Bioengineering, XXVI,* pp. 613–619.

7 CASE STUDIES AND APPLICATIONS

INTRODUCTION

The purpose of this chapter is to illustrate with actual case studies the application of the process analysis techniques that use respirometry for calibration. It is important to remember that the utilization of respirometrically calibrated activated sludge models are usable for both operational and design purposes. The reader should recognize that this technology is applicable for both design and operation.

It is especially useful for operational applications because the collection of respirometric data often takes less than a day; the completion of ancillary modeling analyses requires two hours or less. A plant upset or production needs often pressure management to make decisions rapidly. The performance of a full-blown treatability test using the "conventional" approach generally takes a minimum of a month to complete. The alternative approach using respirometry rapidly identifies a suitable strategy for the operation of the biological treatment facility.

For design applications, respirometric techniques and the associated process modeling do not replace the conventional treatability study; this methodology augments the quantity and quality of data obtained during these efforts. In some cases, experience has shown that these techniques enable one to "fast-track" projects, curtailing both the associated time and the effort required for the conventional approach. When a treatability study is warranted, it entails at the very least the operation of one or more bench-scale reactors. With this level of effort already invested in a project for reactor maintenance, etc., the generation of batch respirometric data is not a significant work addition to a project. During the course of a treatability study, the periodic determination of the biokinetic constants enables the designer to assess both the range and the variation of the values of these parameters. One can also evaluate the time needed for acclimation. With this information and the use of the model, it is feasible to perform a relatively thorough value engineering analysis. A process analyst performs a value engineering analysis by generating a series of predictive curves for effluent quality. Different values of the biokinetic constants generate different predictive curves.

This chapter will present four case studies that are representative examples of applying the respirometric techniques and associated modeling methodologies. Three of the case studies presented in this chapter relate the application and use of the model. Another case study involves the application of respirometry for screening applications. Each case study emphasizes a different type of application and the associated methodology involved with performing the process analysis.

The first case study concerns the use of the respirometric methodologies to perform a relatively lengthy design and operational analysis. This effort involved a municipal treatment facility that encountered difficulties caused by the inhibitory nature of the influent wastewater. The techniques described herein were used to analyze the treatment situation, devise a concept design for a facility's expansion, and perform a verification analysis of the model. Data were collected to compare model predictions and actual values for effluent quality.

Another case study involves the use of these analytical methodologies to determine startup criteria for an activated sludge facility located at a Superfund Site. The facility had to treat wastewaters impounded in several lagoons at the site. This case study is a good example of the ability of this technology to enable a project manager to fast-track a process startup, design, or operational modification. Acute time and budgetary constraints characterized this effort. The respirometric methodology and associated modeling protocol provided accurate information for the startup and operations. The actual operating data for the facility verified the integrity of the respirometric approach.

The third case study involves a biological treatment facility located in a chemical manufacturing plant. Discharges of the heavy metal cobalt to the activated sludge process concerned operations staff. The methodology for process analysis described in this book analyzed this treatment situation and determined the impact of cobalt on process performance. A model analysis for process performance used the data and quantified the impact of cobalt on the activated sludge facility.

A fourth case study will describe the use of respirometry to screen, or rank, the biodegradability of various waste products at a specialty chemical manufacturing facility. The wastewater treatment facility faced the possibility of relatively tight COD limits for effluent quality. The objective of the work presented herein was to "rank" the various waste products produced at the manufacturing facility. These data provide guidance to determine which wastes to remove from the influent waste stream to the biological treatment system. Alternatively, the difficult wastes can receive special consideration for treatment in order to prevent excess pass-through of COD in the plant effluent.

CASE 1: PATAPSCO WASTEWATER TREATMENT PLANT, BALTIMORE, MARYLAND

Background

The Patapsco Wastewater Treatment Plant in Baltimore, Maryland represents a unique treatment situation in that it is a municipal treatment plant receiving a relatively heavy loading of industrial effluents from chemical manufacturers. Additionally, because of concerns over space constraints, the activated sludge system was designed as an oxygen activated sludge facility to conserve land utilization; this means that the design nominal hydraulic detention time (excluding the impact of the recycle stream) is 2 hr. Because of the need to remove phosphorus, the plant was modified to operate using an anaerobic selector as a means to foster a biomass with the capability to remove phosphorus biologically. This process modification means that an aerobic treatment system that already has difficulty with process performance will be further burdened with having one-fourth of the aeration tank (the size of the anaerobic selector section) removed from aerobic treatment capability.

The consulting effort involved both an operations and a design analysis; a design analysis was included because the plant was to undergo an expansion in order to accommodate an increased hydraulic load. Plant operations personnel desired to have the issue of influent waste toxicity addressed as part of both the operations and design analyses. The overall effort consisted of four main tasks:

- Model analysis and review of existing data.
- Determination of biokinetic constants.
- Field evaluation of model predictions.
- Concept design and operational recommendations.

Each of these tasks will be reviewed in detail below.

Model Analysis and Review of Existing Data

This task essentially consisted of a "forensic engineering" effort; the plant had experienced a number of upsets which consisted of gross leakage of COD and five-day BOD (BOD_5); these were attributed to "toxic events" during which large quantities of toxics were presumably discharged to the plant. These occurrences were identified via interviews with plant staff and review of appropriate operating records. The data review showed that the plant did experience large, temporary increases in influent COD (approximately 900 mg/L) associated with these events.

It was decided to utilize the model presented in Chapter 3 to analyze the biological system in an attempt to identify any trends with regard to key operational parameters, the influent loading condition observed during the toxic events, and the occurrences of effluent COD leakage. At this point in the effort, there was no biokinetic data available, so it was decided to analyze the activated sludge process using the model and to use literature values for the biokinetic constants. Since the key parameters that determine effluent quality are the biokinetic growth constants, and since plant personnel were fairly convinced that the upsets were due to the input of toxics, the assumption was made that the waste was inhibitory, and the model analysis was performed using a wide range of values for μ_{max}, K_s, and K_i; the selected ranges for these constants were 0.05 to 0.35 hr^{-1}, 50 to 100 mg/L, and 100 to 300 mg/L, respectively. Since the yield and decay constants, Y_t and k_d, have a relatively small impact on prediction of effluent quality, one set of values, 0.50 mg/mg and 0.01 hr^{-1} respectively, were utilized throughout the analysis.

Another feature that was thought to have an influence over the treatment situation at Patapsco was the fact that there was little or no control over the return sludge flow rate. That is, the return sludge flow rate, F_R, was essentially constant and α, the return flow ratio, was thus allowed to vary. It was determined that this aspect warranted investigation. The return flow issue was investigated by modifying the model presented in Chapter 3, which assumed a constant recycle flow rate in lieu of a constant recycle flow ratio; the "constant F_R" model is produced by substituting F_R/F for α. Comparative process performance predictions using both versions of the model were made for the Patapsco biological treatment system.

Other values for the model analysis were obtained via review of plant operating records or through interviews with plant staff. Since the original plant design specified a 2.0 hr nominal detention time in the aeration tanks, the model analysis evaluated process performance in the range of 0 to 5.0 hr. The impact of influent waste concentration, S_i, was examined over a range of 250 to 1000 mg/L COD. At this point in time, primary influent COD values were around 500 mg/L while shock loads of up to 1000 mg/L were known to occur periodically. The recycle sludge concentration, X_R, was reported to attain concentrations as high as 20,000 mg/L, but a more conservative value of 15,000 mg/L was used for the model analysis. Table 7.1 summarizes the values of all the engineering and biological parameters that were used in the model analysis.

Predictive curves for effluent quality are presented in Figures 7.1, 7.2, 7.3, and 7.4. When the constant α model was used, a value of 0.25 was used for α while a value of 5.0 million gallons per day (MGD) was used for F_R when the constant recycle flow version of the model was used. Both values were selected after interviews with plant staff. Since the reactors have a volume of 1.9 million gallons (MG), an additional abscissa scale is used that presents predictive results as a function of flow rate per reactor (MGD per reactor). Results were presented in this manner so that the required number of reactors that are needed to treat a given influent flow rate can be easily calculated.

The predictive curves that were produced as a result of the model analysis make two important points regarding this treatment situation.

Table 7.1 Values of Engineering and Biokinetic Constants Used for Model Analysis

Biokinetic Constant	Value or Range
μ_{max}, hr^{-1}	0.05 to 0.35
K_s, mg/L COD	50 to 100
K_i, mg/L COD	100 to 300
Y_t, mg/mg	0.5
k_d, hr^{-1}	0.01
Engineering Constants	
S_i, mg/L COD	250 to 1,000
t, hours	0 to 5.0
F_R, MGD	5.0
X_R, mg/L	15,000
V, MG	1.9

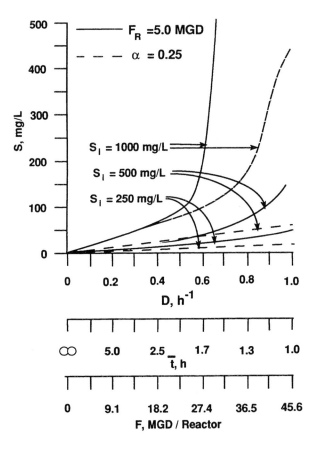

Figure 7.1. Predictive curves for effluent quality showing effect of flow rate and influent waste concentration, S_i, on effluent quality, S, as measured by soluble metabolizable COD, for a 1.9 MG reactor. Values of constants used for the computations were: $X_R = 15,000$ mg/L; $S_R = 30$ mg/L; $\mu_{max} = 0.20$ hr^{-1}; $K_s = 100$ mg/L; $K_i = 300$ mg/L; $Y_t = 0.50$ mg/mg; $k_d = 0.01$ hr^{-1}

First, it is interesting to scrutinize these curves in light of the toxic events, i.e., incidences of increased influent and/or effluent toxicity, that were reported by plant staff. In one case, the plant was subjected to a high-strength (approximately 1,000 mg/L COD), high-toxicity shock. A scan

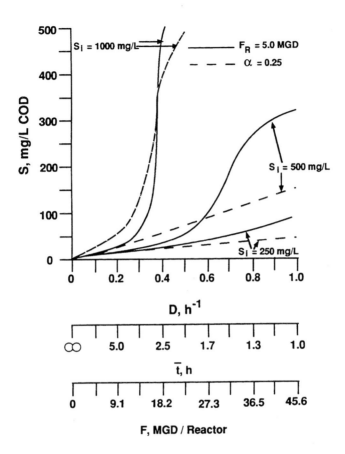

Figure 7.2. Predictive curves for effluent quality showing effect of flow rate and influent waste concentration, S_i, on effluent quality, S, as measured by soluble metabolizable COD, for a 1.9 MG reactor. Values of constants used for the computations were: $X_R = 15,000$ mg/L; $S_R = 30$ mg/L; $\mu_{max} = 0.10$ hr^{-1}; $K_s = 100$ mg/L; $K_i = 300$ mg/L; $Y_t = 0.50$ mg/mg; $k_d = 0.01$ hr^{-1}. Note difference in μ_{max} values compared to Figure 7.1.

of the predictive curves in Figures 7.2 and 7.4 shows that a flow rate of about 16 MGD per reactor at an influent COD concentration of 1,000 mg/L results in a fair amount of effluent deterioration. It should be

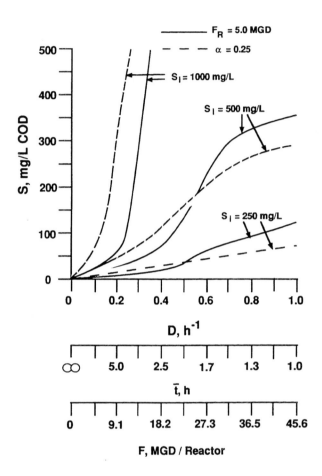

Figure 7.3. Predictive curves for effluent quality showing effect of flow rate and influent waste concentration, S_i, on effluent quality, S, as measured by soluble metabolizable COD, for a 1.9 MG reactor. Values of constants used for the computations were: X_R = 15,000 mg/L; S_R = 30 mg/L; μ_{max} = 0.05 hr^{-1}; K_s = 100 mg/L; K_i = 300 mg/L; Y_t = 0.50 mg/mg; k_d = 0.01 hr^{-1}. Compare with Figures 7.1 and 7.2.

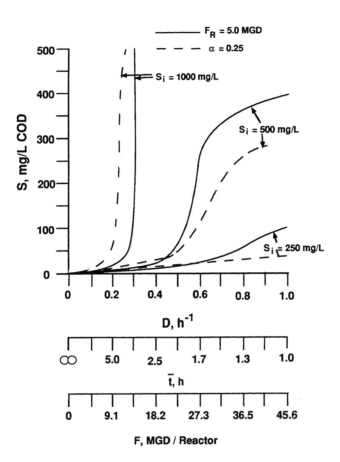

Figure 7.4. Predictive curves for effluent quality showing effect of flow rate and influent waste concentration, S_i, on effluent quality, S, as measured by soluble metabolizable COD, for a 1.9 MG reactor. Values of constants used for the computations were: X_R = 15,000 mg/L; S_R = 30 mg/L; μ_{max} = 0.10 hr^{-1}; K_s = 50 mg/L; K_i = 100 mg/L; Y_t = 0.50 mg/mg; k_d = 0.01 hr^{-1}. Compare with Figures 7.1, 7.2, and 7.3.

noted that although both model analyses in these figures employ a μ_{max} of 0.10 hr^{-1}, the deterioration predicted in Figure 7.4 is worse because these model predictions were made with smaller K_s and K_i values. Data showed that the plant did in fact experience substantial leakage of toxic material during this incident. Thus, it is reasonable to state that the model yielded a reasonably good qualitative prediction (with regard to COD/toxicity leakage) during this time period.

The other important point that the model analysis makes regarding this treatment situation concerns the impact of the recycle parameters on reactor performance. The predictive curves show that the constant F_R operating mode is more sensitive to process upset than the constant recycle flow ratio mode (constant α). This point is readily understood if one examines the equation for net reactor growth rate for the activated sludge reactor.

$$\mu_n = (1 + \alpha + \alpha X_R/X)/t \qquad (7.1)$$

As previously discussed, the constant F_R model has α defined as F_R/F. This means that as detention time, $\bar{t}$, decreases and flow rate, F, increases, the recycle flow ratio, α, is decreasing since F_R is held constant. Reactor growth rate computations using equation 7.1 show that the impact of decreasing α values is to increase μ_n in the reactor; consequently, increases in flow rate increase reactor growth rate by decreasing detention time, $\bar{t}$, and by decreasing α in the constant F_R mode. In contrast, flow rate changes in the constant α mode only impact growth rate by virtue of affecting $\bar{t}$. This is reflected in the prediction of less COD leakage for the constant α mode. The plant eventually implemented a recycle flow system that was "paced" to the influent flow rate (ironically, this operating change was implemented for other reasons); this in effect maintained a constant α or recycle flow ratio. The key point is that the maintenance of a constant α provides protection against effluent deterioration because it enables the operator to impart more control over reactor growth rate. As pointed out in the predictive curves, this is especially the case when a plant is under stress or shock load conditions such as those experienced by Patapsco.

Equation 7.1 also makes an important point regarding the importance of the recycle sludge concentration X_R in maintaining reactor growth rates in the reactors. Since this system has such a low $\bar{t}$ (2 hr for Patapsco, while "conventional" municipal systems have an 8-hr detention time), the bulk of the control over growth rate has to come from the recycle parameters, recycle flow ratio, or recycle flow rate. It was previously noted that a value of 15,000 mg/L was used for X_R in the modeling analyses. Other model analyses, such as that depicted in Figure 7.5, showed that drops in X_R to 10,000 mg/L could adversely impact process

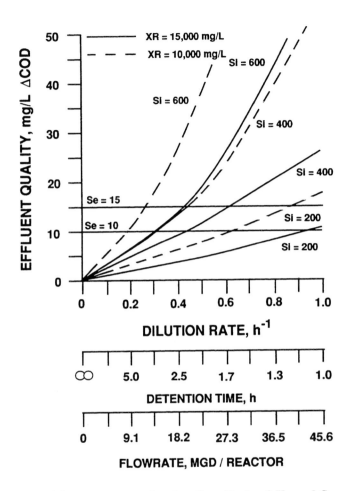

Figure 7.5. Dilute-out curve showing the effects of X_R and S_i on flow capacity of a 1.9 MG activated sludge system. S_e represents different permit limits for effluent quality. Other parameters used in this analysis were $\mu_{max} = 0.108$ hr^{-1}, $K_s = 34$ mg/L ΔCOD, $Y_t = 0.4$, $k_d = 0.005$ hr^{-1}, and $\alpha = 0.25$.

performance; a review of operating data showed that fluctuations of X_R to this level occurred routinely. Poor performance is attributed to the fact that X_R heavily impacts reactor growth rate; this performance problem caused by fluctuating X_R is especially exacerbated when the reactor detention time is on the low end, as is the case with Patapsco. Increases in S_i, the influent waste concentration, serve to stress this situation further, as depicted by the various predictive curves. Control of reactor growth rate is heavily influenced by detention time and the recycle parameters. In the case of Patapsco, the low detention times force a heavy reliance on the recycle parameters to maintain the necessary growth rates that will enable the system to prevent leakage of COD.

Determination of Biokinetic Constants

The biokinetic constants were determined using two different methods: the batch growth study, or shake flask method and the respirometric method. The initial effort for this work focused on utilizing the shake flask method because the respirometric technique had not yet been fully developed. The shake flask method consists of inoculating waste samples at different COD concentrations with the target biomass; the change in biomass concentration over time is then determined using optical density measurements. Once these data are collected, they are analyzed to obtain a set of μ and S points which are in turn analyzed to determine the values of the biokinetic constants, μ_{max}, K_s, and K_i (if the waste is inhibitory). The application of the shake flask technique is usually straightforward when employing synthetic wastes such as those utilized in research efforts. It is quite another matter when the method is applied to a "real" waste stream such as that found at the Patapsco plant.

During the course of work, we noted several interferences that greatly hindered the application of the shake flask technique. One interference was the fact that often the soluble COD of the waste that was available as a substrate source for the biomass was very low; this means that only marginal biomass production can be expected during a test, which greatly hinders the accurate calculation of a growth rate because there will be such a small change in biomass concentration. The only practical countermeasure is to concentrate the waste via an evaporative technique; this was done by evaporating the waste under reduced pressure at relatively low temperatures (55 to 60°C) and adding distilled water to constitute a more concentrated waste; if solids precipitation occurred, the modified waste mixture was filtered. Not only is this procedure labor-intensive, it may compromise the test results if the nature of the substrate is modified as a result of the concentrating method.

Other interferences included the existence of relatively high levels of suspended solids and coloring of the waste, which interfered with absorbance measurements. It was observed that the waste was occasionally colored and the level of coloring was pH-dependent and would change during the course of a test. These interferences made interpretation of the shake flask data time-consuming because analysis of the growth data would often produce equivocal results. Conversely, when we started applying the respirometric method described in Chapter 5, i.e., collecting respirometric data, "translating" it to simulated biomass growth curves, and then calculating growth rates from these curves, we found that the greater majority of our kinetic analyses were extremely straightforward. In point of fact, and employing a popular contemporary expression, we found the analyses of the curves that resulted from the respirometric analyses to be "no brainers."

The results of the initial biokinetic determination efforts for the Patapsco plant are given in Tables 7.2 and 7.3. Table 7.2 presents the values of the biokinetic constants determined using the shake flask method, while Table 7.3 gives the values for the constants that resulted

Table 7.2 Biokinetic Constants from Batch Growth Studies with Patapsco Waste

INHIBITORY RESPONSE

Date	μ_{max} hr^{-1}	K_s mg/L	K_i mg/L	SSQ*
3/6/86	0.149	6	193	$0.43(10^{-3})$
3/13/86	0.222	47	89	$0.20(10^{-2})$
3/18/86	0.296	23	80	$0.30(10^{-2})$
4/16/86	0.279	51	1497	$0.47(10^{-2})$
4/22/86	0.182	63	191	$0.35(10^{-3})$
5/20/86	0.364	71	706	$0.38(10^{-2})$

NONINHIBITORY RESPONSE

Date	μ_{max} hr^{-1}	K_s mg/L	SSQ
3/25/86	0.207	83	$0.69(10^{-3})$
3/27/86	0.217	47	$0.79(10^{-3})$
5/1/86	0.248	34	$0.31(10^{-2})$
5/27/86	0.180	176	$0.10(10^{-2})$

*Sum of squared residuals resulting from data fit.

Table 7.3 Summary of Growth Kinetics Using the Electrolytic Respirometer

| | | Percentage of Waste Strength | | | | | | | | | | μmax | Biokinetic Constants | | |
| | | | | | | | | | | | | | K_s | K_i | SSQ |
Date		15	20	25	40	55	60	70	80	85	100	h^{-1}	mg COD/L	mg COD/L	
3/17/87	μ (h^{-1})		0.063		0.084		0.108		0.113		0.111				
	CODo (mg COD/L)		70		99		121		148		172	0.237	172	—	0.207 E-3
	S$_o$ (mg COD/L)		19		38		56		75		94	0.147	25	—	0.960 E-4
3/24/87	μ (h^{-1})		0.054		0.059		0.048		0.064		0.053				
	CODo (mg COD/L)		68		102		104		170		197	0.060	9	—	0.136 E-3
	S$_o$ (mg COD/L)		25		50		75		100		125	0.057	1	—	0.147 E-3
4/07/87	μ (h^{-1})		0.012		0.019		0.016		0.021		0.021				
	CODo (mg COD/L)		47		70		92		121		153	0.031	68	—	0.110 E-4
	S$_o$ (mg COD/L)		16		32		48		64		80	0.025	16	—	0.140 E-4
4/14/87	μ (h^{-1})		0.036		0.063		0.063		0.075		0.072				
	CODo (mg COD/L)		55		95		121		161		181	0.121	108	—	0.860 E-4
	S$_o$ (mg COD/L)		21		42		64		85		106	0.096	29	—	0.900 E-4
4/27/87	μ (h^{-1})		0.061		0.060		0.078		0.082		0.090				
	CODo (mg COD/L)		39		77		92		105		120	0.113	42	—	0.257 E-3
	S$_o$ (mg COD/L)		7		14		22		29		36	0.099	6	—	0.170 E-3

Date	Parameter										E values
5/12/87	μ (h^{-1})	0.053	0.060	0.112	0.105	0.099	0.273	0.272			0.690 E-3
	CODo (mg COD/L)	61	106	158	234	290	241	379			0.700 E-3
	So (mg COD/L)	28	55	83	110	138	121	197			
7/07/87	μ (h^{-1})	0.034	0.081	0.090	0.100	0.106	0.194	0.171			0.960 E-4
	CODo (mg COD/L)	29	68	99	119	138	138	111	—	—	0.231 E-3
	So (mg COD/L)	16	32	48	64	80	80	45			
7/07/87*	μ (h^{-1})	0.017	0.039	0.064	0.051	0.049	0.097	0.081			0.344 E-3
	CODo (mg COD/L)	33	67	102	119	143	143	86	100	—	0.411 E-3
	So (mg COD/L)	17	34	52	69	86	86	36	36	—	
9/09/87+	μ (h^{-1})	0.319	0.381	0.347	0.275	0.215	0.205	0.321	0.641	0.605	0.140 E-1
	CODo (mg COD/L)	17	28	45	63	88	105	113	11	67	0.140 E-1
	So (mg COD/L)	5	8	13	19	26	31	34	3	22	

*Approximately 52% of the primary solids were removed from the waste sample before the test was started.

+ RAS seed taken from A/O process.

from the respirometric technique described in Chapter 5. It is interesting to note that the values of the constants in both tables are in approximately in the same range. However, except for a few instances, the constants were characteristic of Monod kinetics and not Haldane. That is, except for a few tests that showed inhibitory characteristics, the data indicated that the inhibitory effect thought to be associated with the influent toxicity problem was not present. This presented somewhat of a problem because all other indicators suggested that Patapsco warranted different consideration than one would give to a "typical" municipal treatment facility. The question, then, is how to demonstrate and quantify this aspect of the treatment situation.

The question of the severity of the Patapsco waste and why it should not be considered typical was addressed by comparing the biokinetic results of the Patapsco plant with those obtained for a municipality that did not receive a substantial industrial flow. Peil and Gaudy (1971) evaluated the biokinetic constants for the Stillwater, Oklahoma Wastewater Treatment Plant; this facility received little industrial input, and it was reasonable to state that facility was representative of a typical domestic waste treatment system. The biokinetic constants for this facility were measured as 0.50 hr^{-1} and 63 mg/L for μ_{max} and K_s, respectively (Monod kinetics); Figure 7.6 compares the "worst case" constants (lowest μ_{max}, highest K_s) with the Stillwater constants. Although this figure clearly shows that the Patapsco system, while not inhibited, still nevertheless represents a more difficult treatment scenario than the "typical" situation, it is not clear how this will impact the design of the expanded facilities. This aspect is addressed by comparing predictive curves for effluent quality for both the Stillwater constants and the Patapsco constants.

Figures 7.7 and 7.8 depict predictive curves, which were generated using the model equations given in Chapter 3, the appropriate values of the biokinetic constants, and the values of the engineering constants that were used in the model analyses in Figures 7.1 through 7.4. Figure 7.7 compares the predictive curves generated using an average kinetic response for the Patapsco plant with the predictive curves produced using the Stillwater constants. It is interesting to note that three of the curves predict satisfactory effluent quality over a wide range of detention times; the curve that predicts some potential for effluent deterioration is the one generated using the average Patapsco constants and an influent waste strength, S_i, of 1000 mg/L.

It needs to be emphasized that the only difference between the Patapsco and typical curves is with regard to the values of the biokinetic

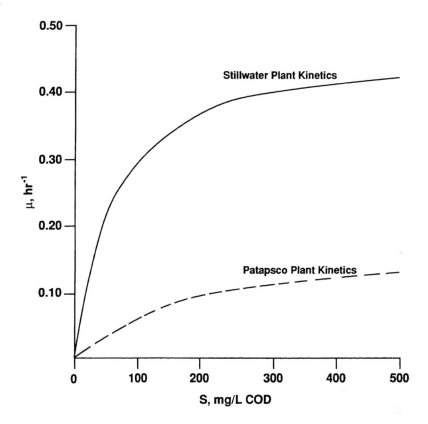

Figure 7.6. Comparison of kinetics for a municipal plant receiving little industrial waste (μ_{max} = 0.50 h^{-1}, K_s = 63 mg/L) and one receiving a heavy industrial waste contribution (μ_{max} = 0.18 hr^{-1}, K_s = mg/L).

constants; that is, any difference in predictions for process performance is solely attributable to the kinetic differences of the two biomass systems. It is also interesting to note that for a wide range of flow rates, the typical system shows little tendency for effluent deterioration, even at an S_i of 1000 mg/L and a detention time of 1.0 hr (at these flow rates, the system would likely fail due to overloading of the secondary clarifier, but not experience a biochemical failure). This result qualitatively agrees with the claims of manufacturers of pure oxygen activated sludge treat-

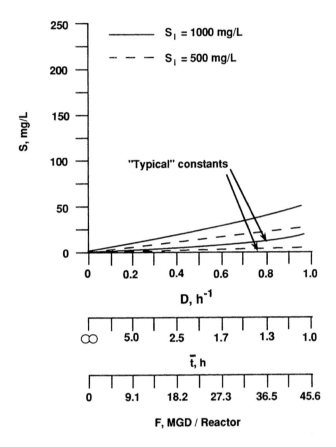

Figure 7.7.   Predicted profile for effluent quality, S, as metabolizable COD for a 1.9 MG reactor. Values of the constants used for computations were: μ_{max} = 0.253 h^{-1}, K$_s$ = 90 mg/L, Y$_t$ = 0.34, k$_d$ = 0.13 d^{-1}, X$_R$ = 15000 mg/L, α = 0.25. "Typical" profile predictions for normal domestic waste generated using μ_{max} = 0.50 h^{-1} and K$_s$ = 63 mg/L (Peil and Gaudy, 1971).

ment systems. That is, it has been stated that these systems can be operated on municipal wastes in high-flow rate, low-detention time situations and deliver acceptable effluent quality (provided that sufficient secondary clarifier capacity is available). If one accepts that the Stillwater con-

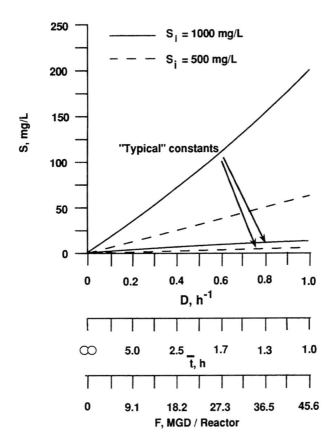

Figure 7.8.   Predicted profile for effluent quality, S, a 1.9 MG reactor for the critical noninhibitory response (low μ_{max}, high K_s). Values of the constants used for computations were: μ_{max} = 0.18 h^{-1}, K_s = 176 mg/L, Y_t = 0.40, k_d = 0.12 d^{-1}, X_R = 15,000 mg/L, α = 0.25. Typical profile predictions for "normal" domestic waste generated using μ_{max} = 0.50 h^{-1} and K_s = 63 mg/L (Peil and Gaudy, 1971).

stants typify the kinetics of municipal activated sludge systems, then the model analysis results provide a reactor engineering argument that supports the high-rate performance contention of proponents of oxygen-activated sludge systems.

When the values of the biokinetic constants are much lower than those of a typical system, Figure 7.8 shows that the situation is more critical. This figure shows that the Patapasco system will have a tendency for substrate leakage at S_i values of both 500 and 1000 mg/L COD, in contrast with the typical system (Stillwater), which shows little tendency for COD leakage over a wide flow rate range. This figure also demonstrates that the biomass kinetics do not have to be inhibitory (Haldane equation) to put a system at risk to experience effluent deterioration caused by COD leakage. In the case of Patapsco, the difficulty with the waste is that the kinetics are low in comparison to a typical municipal waste. Most design bases for municipal systems are based on empirical information. Since the kinetic characteristics of the plant are atypical for a municipal system, then the design basis is also atypical as defined by the values of the biokinetic constants and the associated modeling analyses. The overall situation was further exacerbated by the short detention time in the reactors and the periodic high-strength shock loads which were referred to as "toxic events."

Field Evaluation of Model Predictions

In this phase of the work, an effort was initiated to validate the predictability of the model given in Chapter 3 and the utility of calibrating it (i.e., determining the values of the biokinetic constants) using the respirometric methods described in Chapter 5. A sampling program was undertaken to obtain primary effluent (influent to the activated sludge system) and secondary effluent samples along with samples of activated sludge biomass which were taken from the recycle sludge line. The influent and effluent samples were analyzed for soluble BOD_5 in order to obtain a measurement of S_i and S, respectively, as ΔCOD (BOD_5 can be converted into ΔCOD); the ultimate goal of the exercise was to determine how well the model predicted S (i.e., effluent quality) as measured as soluble ΔCOD or BOD_5.

Using respirometry, nine sets of biokinetic constants were determined from April through September, 1989. On each day that biomass were harvested for testing, influent and effluent samples and appropriate analyses were also taken and performed as previously described. Additionally, the values of the engineering parameters (recycle sludge concentration, X_R, recycle flow ratio, α, and primary effluent flow rate, F) were also recorded on the particular sampling day.

The values of the biokinetic constants obtained as part of this effort are given in Table 7.4; this table shows that only three of the sets of constants were inhibitory. Table 7.5 provides the list of engineering parameters, influent and effluent data, and model predictions. The model predictions were made using the values of the biokinetic constants in Table 7.4, the values of the engineering parameters given in Table 7.5,

Table 7.4 Values of the Biokinetic Constants for the Patapsco Plant Biomass

Date	μ_{max} h^{-1}	K_s mg/L ΔCOD	K_i mg/L ΔCOD	Y_t mg/mg
4/25/89	0.109	15	–	0.41
4/27/89	0.090	15	–	0.41
7/24/89	0.064	28	108	0.29
7/31/89	0.027	17	–	0.29
8/9/89	0.053	28	108	0.31
8/17/89	0.036	16	–	0.27
8/29/89	0.155	1.4	204	0.55
9/5/90	0.090	1	–	0.46
9/11/89	0.037	15	–	0.27

Note: $k_d = 0.005$ d^{-1} was assumed for each study.

Table 7.5 Values of Engineering Parameters and Actual and Predicted Values for Effluent Quality for the Patapsco Plant

Date	X_R mg/L VSS	α	S_i mg/L ΔCOD	F MGD	Soluble Effluent ΔCOD mg/L Actual	Pred.
4/25/89	14,035	0.37	161	45.7	5	1
4/27/89	9,485	0.40	167	45.3	11	2
7/24/89	14,600	0.38	143	55.4	6	3
7/31/89	13,000	0.34	159	73.9	5	9
8/9/89	14,130	0.44	115	44.0	8	2
8/16/89	10,730	0.44	121	44.6	·3	2
8/29/89	9,310	0.37	169	41.5	5	0.1
9/5/89	8,900	0.35	121	40.0	3	0.1
9/11/89	10,360	0.37	192	44.4	6	4

and the model equations provided in Chapter 3. Table 7.5 shows that predicted and actual values for effluent quality were relatively reasonable, especially considering the complex nature of trying to predict effluent quality in a large (70 MGD) operating facility. The difficulty encountered in trying to predict low values of a given parameter should also be emphasized. This also supports the conclusion that the model calibrated with biokinetic constants determined via analyses of respirometric data provided reasonable predictions of plant effluent quality.

Given the favorable predictions provided by the model and the supporting positive operational data, one may question the utility of performing the modeling exercise. The answer involves the need to predict the operating envelope (those operating conditions that could put the plant at risk of effluent deterioration and subsequential permit violation). Figures 7.9 and 7.10 illustrate the utility of the modeling approach in predicting plant capacity for various waste strengths (S_i). The predictive curves in Figure 7.9 were constructed utilizing the kinetic constants and engineering parameters that were relevant for 4/27/89, while those curves for Figure 7.10 were generated using the appropriate information from 8/17/89. The figures show that, at the actual S_i values measured for each day (167 mg/L and 121 mg/L for 4/27/89 and 8/17/89, respectively), the plant has a wide operating range before experiencing problems; as previously discussed, the actual field data showed that the plant had little trouble in delivering excellent effluent quality at these S_i values. (In this case, plant capacity is more likely to be limited by the capacity of the secondary clarifiers as influent flow rate increases and detention time

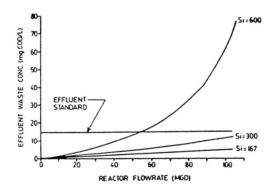

Figure 7.9.   Predictive curves for effluent quality for the Patapsco plant (4/27/89) from Colvin et al. (1991).

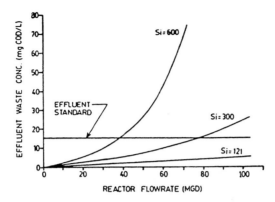

Figure 7.10. Predictive curves for effluent quality for the Patapsco plant (8/17/89) from Colvin et al. (1991).

decreases.) Both figures indicate that as S_i increases to 600 mg/L COD, the reactors are susceptible to effluent deterioration via COD leakage. If one utilizes 15 mg/L as effluent criteria (see Colvin et al. (1991) for a detailed explanation), the plant capacity at an S_i of 600 mg/L is 55 MGD for 4/27/89 and 38 MGD for 8/17/89.

It should be noted that the primary difference in conditions between the predictions in Figures 7.9 and 7.10 is the difference in values between the values of the biokinetic constant, μ_{max}. The engineering parameters used for the predictions are relatively close in value. The point is that the model predictions are sensitive to the values of the biokinetic constants determined using analyses of respirometric data via the methodology described in Chapter 5. The biokinetic constants, in turn, are sensitive to various environmental and plant conditions (e.g., waste quality, temperature, pH, etc.) that affect the ecology of the activated sludge biomass; these aspects were discussed more fully in Chapter 6. Thus, as conditions change that affect the biomass, and consequently plant capacity, the modeling procedure (when updated with respirometric calibration to determine new values of the biokinetic constants) has the ability to quantify the impact of environmental changes on plant capacity.

Operational and Design Recommendations

The key question to be answered at this point concerns the recommendations made to the City of Baltimore that resulted from the analytical and modeling efforts. Two primary recommendations were made:

- Additional aeration volume to meet the expanded treatment needs.
- Additional level of control and operational flexibility for the biological system in view of the difficult nature of the waste.

Based on the design analysis as exemplified by the predictive curves for effluent quality given in Figures 7.7 and 7.8, it was decided that the 1.9 MG reactor could safely handle a flow rate of approximately 18.2 MGD. It was also noted that each reactor will only utilize 75% of its volume for aerobic treatment since the first quarter of the reactor was to be held anaerobic in order to provide a selector mechanism to promote biological phosphorus removal. Using these numbers and a design flow of 87.5 MGD, one computes a need for 6.4 reactors ($87.5/(18.2\cdot0.75)$); this calculation suggests the need for three additional reactors because the plant already has four. This suggestion created a problem with some of the design engineering staff because the original recommendation called for only one additional reactor. However, this recommendation was based on the assumption that the Patapsco waste was a conventional municipal waste. Clearly, the kinetic results (e.g., refer again to Figure 7.6, which compares the Patapsco kinetic results to those of a more "conventional" municipal waste) and associated modeling analyses showed that this was not the case. However, an increase from one to three reactors was still a bit hard to swallow for the design engineering staff. We then suggested that, if the plant were given some level of operational flexibility and control for the recycle parameters (recycle sludge concentration X_R and recycle flow ratio α) then two additional 1.9 MG reactors would be acceptable because control of X_R would enable the plant to "buy" more capacity by virtue of the fact that this control technique would enable the plant to increase its operating envelope. (It should be noted that as part of the retrofit effort to implement biological phosphorus removal, the plant had control over recycle flow rate which enabled it to maintain a constant α.) Thus, if conditions changed, X_R could be increased to increase capacity. It should be stressed that the design already called for a short detention time; this situation was further exacerbated by the conversion of 25% of the activated sludge reactor volume to an anaerobic

zone. Given the difficult nature of the waste and the other considerations, the recommendation of at least some form of X_R control or the
provision of some means to bolster recycle sludge concentration should
be viewed as essentially a basic need for this system.

Figure 7.11 shows an example of the application of recycle sludge
concentration control for an activated sludge plant. (The example shown
is not for Patapsco because, at the time of the preparation of this book,
other considerations precluded the incorporation of all of the recommendations into the final design.) The concept design shown in Figure 7.11 is
for an industrial waste treatment system that must handle high-strength
organic wastes, which are characterized by inhibition. One key feature
for the design is the use of thickeners to provide additional thickening
capability for the return activated sludge, X_R; it may also be possible to
realize this goal by simply utilizing polymer dosing to the mixed liquor to
increase underflow sludge concentration. Another feature is the use of
dosing tanks to store the thickened return sludge and to maintain X_R at a
constant concentration. It should be noted that these tanks will also
function to store excess biomass in case the system takes an inadvertent
toxic shock load or "hit." The excess biomass storage capacity also
enables plant personnel to bring up an out-of-service aeration tank rapidly in case there is an unforeseen shift in production schedules that
causes an increase in plant loading. If there is a change in the quality or
quantity of the waste that causes a change in the values of the biokinetic
constants, then plant personnel have the flexibility to reevaluate the

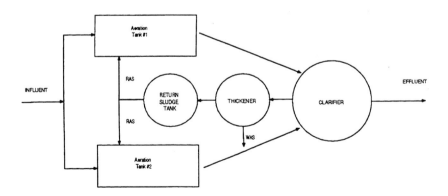

**Figure 7.11.   Concept design for activated sludge system using control
of X_R.**

constants (respirometrically, of course), insert these values and the projected loading information into the model, and then use the model to determine the recycle values needed to maintain effluent quality; the plant design given in Figure 7.11 will then enable the operators to "dial-in" the appropriate recycle values which are needed to keep the plant in compliance.

CASE 2: STARTUP AND OPERATION OF BIOLOGICAL TREATMENT FACILITY FOR IMPOUNDMENT LEACHATES AND WASTEWATERS AT A SUPERFUND SITE

Background

A potentially responsible party (PRP) was charged with having to implement the treatment of impoundment leachates at a Superfund site. The leachates were primarily contaminated with organics produced at the manufacturing facility that used to be operated at the site. The manufacturing facility associated with the site produced rayon, polyester, and polypropylene fibers; however, the primary product was rayon and filament fiber.

As a result of the production of these fibers, several waste products were generated and disposed in surface impoundments. These wastes included viscose liquid and solids, primary metal precipitates, waste activated sludge from the plant wastewater treatment system, and coal combustion flyash from the facility boiler room. Most of the liquid and solid wastes were conveyed to surface impoundments and landfill areas. Eleven earthen impoundments contained viscose, which is a waste material which is generated from the backwashing of viscose filters used to remove undissolved cellulose particles and waste associated with the fiber-making process. The other basins primarily held inorganic wastes, waste activated sludge and flyash.

The large impoundments were used to collect stormwater and other runoff waters. The regulatory agency that was monitoring site cleanup activities noticed that the impoundments were essentially filled to capacity. It was also noted that a heavy rainstorm could cause the impoundment which collected leachates, wastewaters, and stormwaters to spill over into river adjacent to the site, which was used for recreational activities and fishing. The situation prompted the oversite agency to compel the PRP to begin treatment of the impoundment liquors in order to keep the liquid level in these waste basins at a manageable level that would not present a threat to the river.

Although analyses of the leachates indicated a fairly high BOD_5, the agency recommended the application of physical/chemical treatment because of the perception that this represented a reliable technological choice. The recommended treatment flow scheme consisted of chemical precipitation, pressure sand filtration, and granular activated carbon. Operating experience showed that despite best efforts, this treatment scheme was unable to meet applicable or relevent and appropriate requirement (ARAR) limits for BOD_5. The oversite agency then required the implementation of biological treatment in order to meet the BOD removal requirements. The PRP decided to restart and use an existing but idle wastewater treatment facility at the site. This facility consisted of primary treatment followed by an activated sludge system. The initial treatment strategy called for pH adjustment (raise the pH) of the impoundment liquor prior to primary treatment to drop out solids that might interfere with the biological treatment process. After primary treatment, the pH was then lowered using acid and the liquor fed to the activated sludge system.

Approach

An initial investigation showed that the activated sludge equipment was in good condition. However, the PRP group was faced with a rigorous compliance deadline and there was little time to perform any sort of involved treatability study to verify the efficacy of biological treatment. This also meant that the PRP would not be able to utilize conventional treatability methods to determine the operating envelope or to optimize operational parameters, which prompted the PRP to employ respirometry testing and associated modeling analyses to determine the plant performance envelope for startup and operations.

The plan for implementing biological treatment of the impoundment liquors involved the use of biomass from a local publicly owned treatment works (POTW) to seed the aeration basins. A series of tests were performed using the methodology described in Chapter 5 to determine the values of the biokinetic constants. A modeling analysis was then performed using the values of the constants, the projected flow rate for the waste F, a range of values for the influent waste strength S_i, and the predictive model presented in Chapter 3.

A graph which provides the initial kinetic results for the impoundment wastes is given in Figure 7.12. Biomass and waste from the site were shipped via an overnite express shipping service to a laboratory for respirometric analyses. The kinetic results and associated modeling analyses were presented to the site wastewater treatment manager the next afternoon (turn-around time of 30 hr). The kinetic results for the startup condition were 0.095 hr1, 62 mg/L, and 351 mg/L for μ_{max}, K_s, and K_i, respectively. Prior to startup, the wastewater treatment management staff wanted to operated the facility based on a target value for mixed liquor of 4,000 mg/L. The modeling analyses compute the value for mixed liquor suspended solids (MLSS) that one can expect. This aspect was especially significant for this facility because the facility was designed for a flow rate of 3500 gallons per minute (gpm) and the flow rate for the impoundment leachate was projected at 400 gpm with a strength of 400 mg/L COD; this translated to a nominal hydraulic detention time of 64 hr. (That is, the activated sludge system will operate in an underloaded condition.)

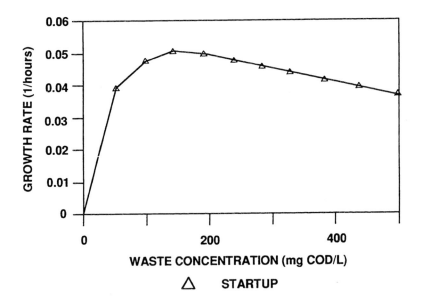

Figure 7.12.   Initial growth kinetics of biomass treating Superfund leachate.

Table 7.6 shows the predicted results for the activated sludge system. Values for the engineering parameters were selected based on best available information for the system; α was 1.0, X_R was 1,000 mg/L, S_i was 400 mg/L COD, and the tank volume, V, was 1.53 MG. The computations given in Table 7.6 predict that the facility will not have a problem meeting effluent requirements (15 mg/L based on COD) over a large operating range. This is due to the large detention time that is available because the tank is underloaded; large detention times place less emphasis on the recycle parameters to keep low system growth rates or high Θ_c values (refer to Equation 4.6). The predictions also show that the reactor can expect to have a mixed liquor volatile suspended solids (MLVSS) value of approximately 600 mg/L; this translates to a projected MLSS value of 860 mg/L assuming a volatile fraction of 70%. This value contrasts to the MLSS value of 4,000 mg/L considered to be reasonable by the facility management staff. The difference can be reconciled by noting the F:M and Θ_c values that were predicted by the model analysis. (It should be noted that the model presented in Chapter 3 is easily modified to present results using parameters for activated sludge that are more familiar to process analysts; a more detailed discussion concerning this topic is given in Chapter 4.) At detention times greater than 83 hr, the model predicts that the system is already beyond the point of extended aeration; that is, X_w, the predicted mass of waste sludge, is negative. This means that the system cannot support the imposed recycle condition of 1,000 mg/L for X_R. Consequently, it is inappropriate to expect that the system can support the higher recycle values (approximately 10,000

Table 7.6 Predictive Model Analyses for StartUp of Superfund Leachate Treatment Facility

t (h)	S mg/L COD	X mg/L VSS	MCRT d	F:M mg COD/mgX/d	X_w lb/d
125	2	540	∞	0.14	<0
100	3	550	∞	0.17	<0
83	3	562	63	0.20	113
63	4	573	20	0.27	365
45	6	582	10	0.36	748
26	11	592	4	0.62	1770

Note: X_R = 1,000 mg/L, V = 1.53 MG, S_i = 400 mg/L COD, μ_{max} = 0.095 hr^{-1}, K_S = 62 mg/L, K_i = 351 mg/L, Y_t = 0.55, and k_d = 0.002 hr^{-1}.

mg/L) that one would expect at the design flow rate of 1600 gpm. At a detention time of 63 hr, Table 7.6 shows that the reactor will operate at a Θ_c of 20 d and an F:M of 0.27 kg COD/kg MLVSS. Although conventional wisdom suggests that these values are adequate to meet target treatment goals, they had to be determined using the "unconventional" modeling approach presented in Chapter 3. That is, the idea of operating the system at an MLSS value of 4,000 mg/L is irrelevant for this treatment situation.

The reactor started up with no problem and the field results indicated that the predictive modeling results were very accurate. Subsequent respirometric tests were performed to refine operations. One set of tests focussed on optimization of feed pH. Other respirometric tests were employed to evaluate the benefits of pretreating the liquor in the primaries to remove suspended solids. A good example of the application of respirometric techniques for process optimization is provided in Figure 7.13. The impact of pH on process kinetics (μ versus S) is depicted in this figure. Raising the pH prior to primary treatment and then bringing it

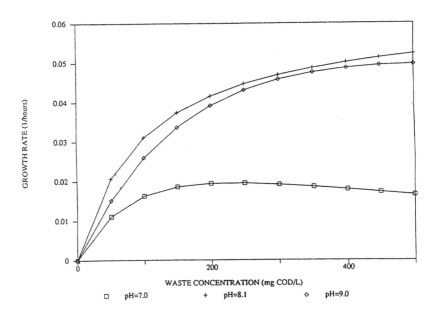

Figure 7.13. Impact of pH on growth kinetics of biomass treating Superfund leachate.

back to neutrality afterwards, presumably to enhance biotreatment operations, resulted in substantial chemical usage. The kinetic analyses in
Figure 7.13 show that adjusting the pH of the reactor feed close to
neutrality is actually detrimental to process kinetics and hinders biomass
performance. Using less acid and keeping pH values closer to those of
the primary effluent saves on chemical usage and enhances biomass
performance. This also serves to improve the process performance envelope of the activated sludge system.

CASE 3: IMPACT OF COBALT ON BIOLOGICAL WASTEWATER TREATMENT PLANT PERFORMANCE

Background

An organic chemicals manufacturing plant consisting of integrated
synthetic fibers and polymer intermediates production facilities discharged wastewaters containing cobalt. The source of the cobalt was
likely attributable to a cobalt/manganese organic acid production catalyst. The wastewater treatment facility consists of an activated sludge
system operated in two stages. The wastewater for the polymer intermediate production facilities is discharged after equalization into the first
stage activated sludge system. This system typically removes about 95%
of the organic acid COD. The effluent from the first system, along with
wastewater from the synthetic fiber manufacturing unit, are discharged
to the second-stage activated sludge system.

Production process modifications were made to comply with a new air
emissions permit. Soon after these productions changes were made,
operations staff noticed a reduction in the treatment efficiency of the
first-stage activated sludge unit. The first stage typically treats 0.5 MGD
of a mildly acidic (pH about 4–5) wastewater which has a high-strength
COD of 10,000 to 15,000 mg/L. Analysis of wastewater feed samples for
the first stage were performed to search for potential inhibitory agents.
The results indicated elevated levels of cobalt and manganese. A review
of the production units showed that the source of the cobalt was likely
the cobalt/manganese catalyst. The major source of catalyst discharge to
the first-stage activated sludge system was believed to be a solids slurry
that was temporarily diverted to wastewater treatment in order to maintain production while process modifications to production units were
made in order to comply with the previously mentioned air permit

requirements. The concentrations of soluble cobalt in the equalization basins were measured at levels from 10 to 40 mg/L after the slurry was introduced to the treatment system.

The general inhibitory effects of metals on activated sludge biomass were well-known. However, specific data regarding the impact of cobalt were not. A testing regimen for kinetic analysis using respirometry was devised in order to quantify the impact of cobalt on plant operations. The conceptual plan was relatively straightforward. The kinetics of activated sludge biomass degrading the first-stage wastewater were evaluated at various cobalt concentrations. The testing methodology is described in Chapter 5, while Chapter 6 discusses conceptual aspects of the impact of toxicants (in this case, cobalt) on growth kinetics (specifically, refer to Figure 6.14). The kinetic effects are quantified by determining the effects of cobalt on μ_{max}, K_s, and K_i (if substrate inhibition is prevalent in the base case, i.e., wastewater free of toxicants). The constants are then inserted into the process model presented in Chapter 3 to quantify the impact of the cobalt on the performance envelope of the first-stage activated sludge system.

Approach

The goal of the kinetic evaluation effort was to evaluate the detrimental impact of cobalt on the performance of the first-stage activated sludge system. Chemical analysis of the first-stage biomass showed that samples had elevated concentrations of cobalt and manganese. Since the goal of the work was to evaluate the impact of cobalt on activated sludge performance, it was decided that the kinetic testing effort required a cobalt-free biomass. A biomass without cobalt contamination but that was acclimated to the organic composition in the first-stage wastewater was needed to develop the "baseline" case. This biomass was developed by obtaining a sample of first-stage biomass that contained cobalt. The biomass was then acclimated to a synthetic wastewater that simulated the first-stage wastewater but did not contain cobalt. The "de-cobalted" biomass was grown in bench-scale activated sludge units with internal recycle (i.e., "Eckenfelder units"). Sludge in the reactor was analyzed on a weekly basis for cobalt concentration. Kinetic testing of the biomass began once the cobalt concentration in the biomass was less than 1 ppm.

A corollary effort involved the assessment of the effectiveness of a pretreatment step to remove cobalt from the influent of the first-stage system. Plant wastewater was pretreated to remove cobalt. Kinetic tests were then performed on raw and pretreated wastewaters using the biomass which was free of cobalt and cultivated in the bench-scale unit.

The numerical results for the biokinetic constants for the cobalt testing are listed in Table 7.7. Comparative biokinetic growth curves are presented in Figures 7.14 and 7.15. Figure 7.14 illustrates the impact of increasing concentrations of cobalt on the biodegradation kinetics of the seed biomass treating the synthetic plant feed. Figure 7.15 shows a com-

Table 7.7 Kinetic Results for Cobalt Wastewater Analysis

Wastewater	μ_{max} h^{-1}	K_S mg/L COD	K_i mg/L COD
Synthetic Feed	0.129	296	970
Synthetic Feed + 1 mg/L Cobalt	0.061	154	685
Synthetic Feed + 10 mg/L Cobalt	0.028	287	1249
Normal Plant Feed	0.068	251	693
Pretreated Plant Feed	0.099	248	1241

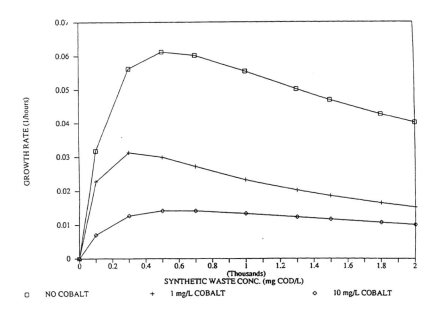

Figure 7.14.   Impact of cobalt concentration on biomass kinetics.

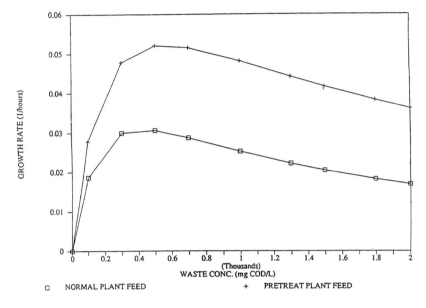

Figure 7.15.   Impact of pretreatment on kinetics of biomass treating cobalt-containing wastewaters.

parison of the biomass kinetics on the normal and pretreated wastewaters. It should be noted that the base case of no cobalt in Figure 7.14 and the pretreated case in Figure 7.15 compare quite favorably. The biomass used for these two tests was the same, but the feeds were different. The test in Figure 7.14 used the synthetic feed while the one in Figure 7.15 employed the pretreated plant wastewater. The fact that the kinetic results are relatively close suggests that the synthetic feed did a good job of mimicking the plant wastewater.

Figure 7.14 makes a number of points. It demonstrates and quantifies the impact of increasing concentrations of cobalt on the biodegradative capability of the biomass. Cobalt clearly has a significant impact on the biomass to degrade the target waste. The biokinetic constants which are listed in Table 7.7 were employed in a process analysis of the first-stage activated sludge system. The process analysis was performed using the predictive equations presented in Chapter 3. Predictive curves for effluent quality are presented in Figures 7.16 and 7.17. The engineering and influent parameters used for these analyses are a recycle sludge concen-

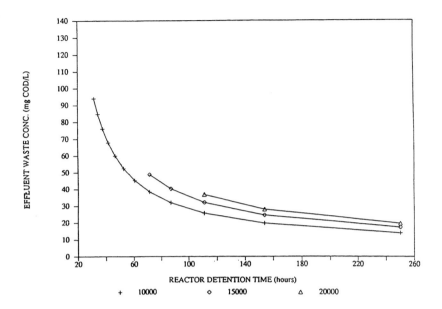

Figure 7.16.   **Predictive curves for effluent quality at different S; values for biomass growing on synthetic wastewater (no cobalt).**

tration, (X_R), of 6,000 mg/L, a recycle ratio (α) of 1.0, and influent COD concentrations $(S_i s)$ of 10,000, 15,000, and 20,000 mg/L.

Figure 7.16 shows that the plant has substantial capacity for treating the synthetic waste. At an S_i of 10,000 mg/L COD, it can process down to a detention time of 40 hr without significant deterioration in effluent quality. Since the waste is inhibitory and the influent strength is high, critical point violations are likely for this waste treatment situation. (For more discussion, please refer to the inhibitory kinetics sections of Chapters 1, 2, and 3.) At 15,000 and 20,000 mg/L influent COD, the limiting detention times are approximately 70 and 110 hr, respectively. Figure 7.17 shows that the impact of 1 mg/L of cobalt is to reduce treatment capacity at an S_i of 10,000 mg/L from a detention time of 40 hr to 70 hr. There is virtually no capacity at 15,000 mg/L COD. Other process analyses of the 10 mg/L cobalt situation showed that there was essentially little processing capability by the biomass. It is recognized that the seed biomass employed in these tests was relatively unacclimated to the presence of cobalt and that a better kinetic response would have been obtained

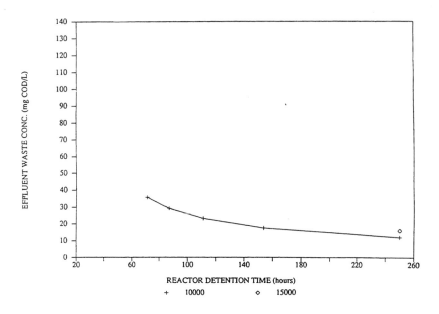

Figure 7.17. **Impact of 1 mg/L cobalt on operating envelope of activated sludge system. Compare with Figure 7.16.**

with a biomass more acclimated to cobalt. However, the analysis showed that eliminating cobalt from the waste stream substantially enhances the performance envelope of the plant. Evaluation of the impact of acclimating the biomass to cobalt by examining "intermediate" levels of cobalt contamination of the biomass as judged to be too time-consuming for the purposes of this work.

Figures 7.18 and 7.19 show favorable results regarding the impact of pretreating the first-stage feed. The biokinetic constants used to generate these curves were the normal plant feed and the pretreated plant feed for Figures 7.18 and 7.19, respectively. The engineering constants were the same as those used in Figures 7.16 and 7.17. A comparison of Figures 7.18 and 7.19 shows that the effect of pretreatment is to extend the performance envelope of the plant. Figure 7.19 shows that the effect of pretreatment is to increase the plant capacity to a detention time of 30 hr at an influent COD of 10,000 mg/L. The limiting detention times for 15,000 mg/L and 20,000 mg/L are 60 and 90 hr, respectively. The normal plant feed has limiting detention times of 80 and 160 hr for S_i values of

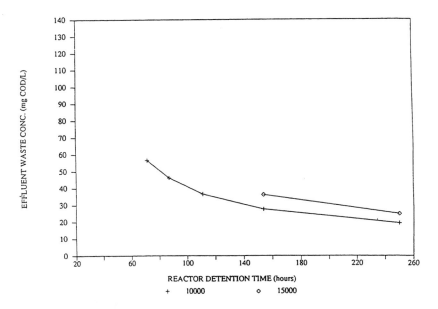

Figure 7.18.   Operating envelope of activated sludge system treating cobalt-containing wastewater.

10,000 mg/L and 15,000 mg/L, respectively. The analyses also showed that at 20,000 mg/L COD, the plant is only able to provide marginal treatment at best for the untreated waste.

This example provides a good case study of how to use respirometric testing and associated modeling analyses to predict and to quantify the impacts of upstream process changes (in this case, pretreatment to remove cobalt) on the performance capacity of the plant. This enables plant operations staff and management to make informed economic decisions regarding operational modifications instead of having to rely on "plant folklore" or less structured analytical approaches.

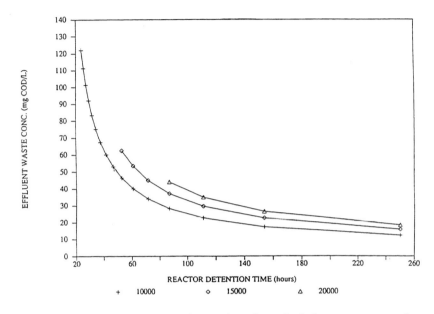

Figure 7.19.   Operating envelope of activated sludge system treating pretreated (cobalt precipitated out) wastewater. Compare with Figure 7.18.

CASE 4: USE OF RESPIROMETRY FOR SCREENING AND RANKING APPLICATIONS

Background

A large international corporation was investigating the biodegradability of five organic chemical products that were soon to be manufactured at an overseas location. The overseas plant was facing the possibility of meeting strict effluent requirements for COD; consequently, it was important to identify any troublesome or recalcitrant products which may pass through the biological treatment process and contribute to effluent COD. A corollary effort involved devising a ranking procedure; both the screening and ranking procedures involved the interpretation of respirometric data.

Approach

The general testing approach involved the operation of a bench-scale activated sludge unit that supplied seed for separate batch testing using respirometry. A schematic outlining this procedure is given in Figure 7.20. A waste stream containing equal-COD portions of the target components was fed to the bench-scale activated sludge reactor. The bench-scale unit operated at a long mean cell residence time in order to simulate extended aeration conditions that would be utilized by the field system. After an acclimation period of three weeks, biomass was harvested from the bench-scale reactor and used in separate batch respirometry tests to identify the biodegradability of the individual component wastes.

A total of four respirometry tests were performed using a six-unit electrolytic respirometer. Each test was performed using six 1-L flasks, with each flask containing the desired amount of waste sample, nutrients, and biomass seed for each test. Cumulative oxygen uptake measurements were taken for a period of 48 hr. The waste samples added to

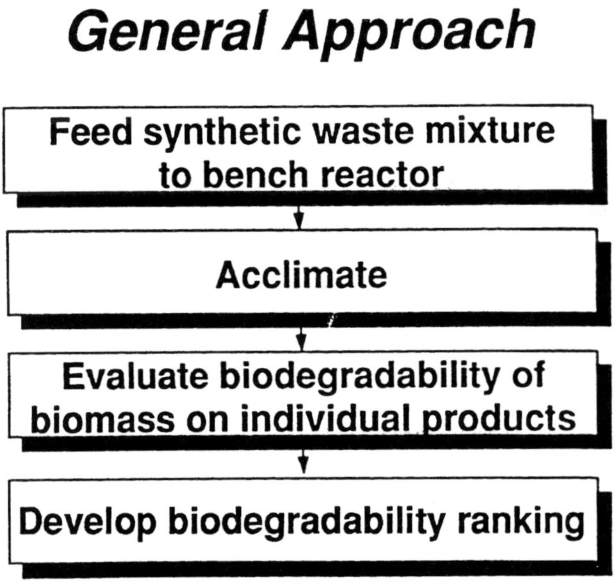

General Approach

Feed synthetic waste mixture to bench reactor

Acclimate

Evaluate biodegradability of biomass on individual products

Develop biodegradability ranking

Figure 7.20. General approach for organic screening procedure using respirometry.

each flask were taken from stock solutions made from concentrated waste samples. Variations in initial COD values in the test flasks were attributed largely to the emulsified nature of the wastes. This points out the utility of respirometry for evaluating biodegradation in this waste treatment situation. Relying on COD measurements alone leads to ambiguous results. The respirometric data augments conventional treatability data and allow the analyst to obtain a clearer interpretation of the biodegradation data.

Results

The results of the batch respirometry tests are depicted in Figures 7.21 through 7.24. Analysis of the data indicated that approximately half of the batch test results represent good COD depletion and O_2 uptake balances. The COD depletion is accounted for in the mass of oxygen utilized and the increase in cell mass. This balance analysis provided an addi-

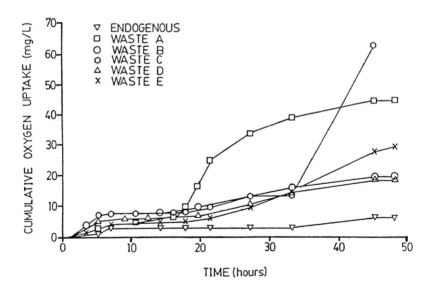

Figure 7.21. Case Study 4 Respirometric Test #1: Individual components plus endogenous seed reactor (Colvin et al., 1991).

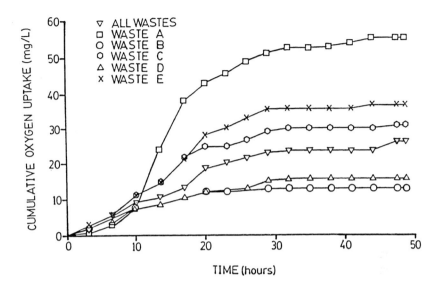

Figure 7.22. Case Study 4 Respirometric Test #2: Individual components plus "All Waste" reactor (Colvin et al., 1991).

tional check on the biodegradation data and aided in eliminating confusing results.

The results of the first respirometry test are presented in Figure 7.21. After a review of the data, it was determined that Waste A was most biodegradable, while Wastes B and D were problematic. The oxygen uptake results for Waste C were judged to be erroneous and inordinately high because the COD depletion did not match O_2 uptake. A check of the equipment indicated that the respirometer had a faulty seal that resulted in an overproduction of oxygen in the Waste C flask. The balance analysis enabled us to sort out the analytical confusion quickly.

The results of the other tests supported the results presented in Figure 7.21. Wastes B and D were problematic; Waste A tested relatively biodegradable with the other wastes remaining somewhere in the middle. These results provided a qualitative ranking of the wastes. The client requested that we provide a more quantitative interpretation of the data. The ultimate goal is to formulate a methodology to to quantify the biodegradation ranking of each waste component and to develop an algorithm for estimating the effluent soluble COD contribution of each

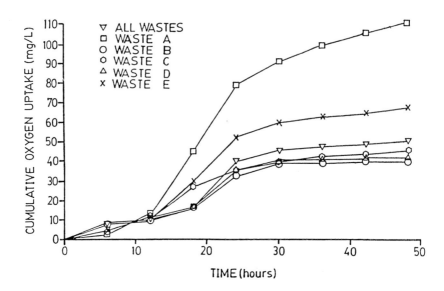

Figure 7.23. Case Study 4 Respirometric Test #3: Individual components plus "All Waste" reactor (Colvin et al., 1991).

waste. The objective is to predict the impact of the individual wastes on effluent quality as measured by COD. The approach utilized the data in the batch test to rank the components by the product of ΔCOD (COD removed) divided by the initial COD used in the test. This gives the fraction of initial COD removed. ΔCOD was computed using the respirometric data according to the following expression:

$$\Delta COD = O_2 \text{ uptake}/(1 - O_x Y_t) \qquad (7.2)$$

Table 7.8 lists the ΔCOD/initial COD values for the five wastes tested. The information in Table 7.8 is used to develop the biodegradability ranking in Table 7.9. In order to predict the impact of the influent waste mixture on effluent COD characteristics, the batch results listed in Table 7.9 need to be "scaled" to the bench-scale system. This is stated because bench-scale reactors that simulate full-scale systems generally have greater COD removal than batch reactors. The bench scale-reactor for this study achieved a 73% removal of COD. The ranking procedure for individual wastes is formulated by assuming that each waste is characterized by its own COD removal ratio. This removal efficiency is assumed

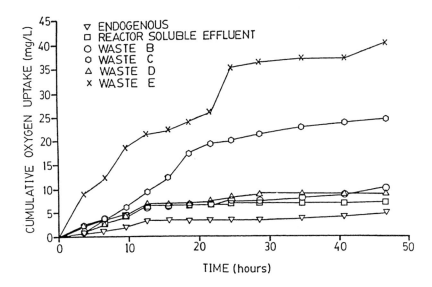

Figure 7.24. Case Study 4 Respirometric Test #4: Components B, C, D, and E: effluent BOD; and endogenous reactor (Colvin et al., 1991).

Table 7.8 ΔCOD/Initial COD Values Calculated for Each Waste from the First Three Batch Respirometry Tests (From Colvin et al., 1991).

Test No.	Waste A	Waste B	Waste C	Waste D	Waste E
1	− − +	0.322	0.361*	0.373	0.713
2	0.561	0.420	0.332	0.139	0.378
3	0.635*	0.220	0.510*	0.790*	0.532
Avg.	0.598	0.321	0.401	0.434	0.541

+ ΔCOD/Initial COD Value was greater than 1.0.
* Used actual final ΔCOD for this computation.
Note: The results of Test 4 are not included because Waste A was not analyzed in this test.

**Table 7.9 Summary of ΔCOD/Initial COD Data and Biodegradability
(From Colvin et al., 1991)**

Waste	Average ΔCOD/Initial COD*
A	0.598
E	0.541
D	0.434
C	0.401
B	0.321

*Test 4 not included in calculations.

to be additive which results in an aggregate ΔCOD for the waste mixture:

$$\Delta COD_A + \Delta COD_B + etc. = \Delta COD_{Waste\ Mixture} \qquad (7.3)$$

Adding up the average ΔCOD/initial COD numbers from Table 7.7 and dividing by 5 (the bench-scale reactor was fed a mixture consisting of five equal-COD contributions of the five waste products) yields 0.459. The bench-scale unit had a COD removal efficiency of 73%. The scale factor is calculated as 0.73/0.459 which equals 1.59. The adjusted ΔCOD/initial COD values for the wastes are given in Table 7.10. The information in Table 7.10 is utilized by inserting each removal factor into a formula for predicting the impact of waste stream composition on effluent COD. It is recognized that this approach is subject to more rigorous evaluation. It does, however, represent a reasonable starting point for utilizing bench-scale respirometric data for predicting the impact of different waste components on effluent COD.

**Table 7.10 Adjusted ΔCOD/Initial COD Factors for Each Waste
(From Colvin et al., 1991)**

Waste	Adjusted Factor
A	0.951
E	0.860
D	0.690
C	0.638
B	0.510
Combined Waste	0.730

KEY CONCEPT SUMMARY

The key concepts contained in this chapter are:

- The process modeling methodology described herein is suitable for both design and operational application. Case 1 presented an example of a design application while Cases 2 and 3 are examples of operational analyses.
- A process analysis is performed by utilizing the predictive equations presented in Chapter 3. If a system does not conform to complete/mix or cells-in-series, then the same derivation procedure presented in Chapter 3 can be used to derive equations which are appropriate for the particular system configuration. The modeling procedure described in Chapter 3 can be applied to any reactor configuration to yield configuration-specific predictive equations.
- The model equations are calibrated using the respirometric techniques given in Chapter 5. Various issues concerning the acquisition of kinetic data and the impact of environmental conditions on the values of the biokinetic constants are discussed in Chapters 5 and 6.
- A model analysis identifies the values of the engineering control parameters wich are needed to prevent plant upset. Alternatively, the analysis quantifies the operating envelope for a given set of influent conditions and values of the biokinetic constants.
- Respirometric methods can also be applied in a screening mode to evaluate the relative potential of wastes to pass through a biological treatment system. These applications are described in Case 4.

REFERENCES AND SUGGESTED ADDITIONAL READING

Colvin, R.J., Rozich, A.F., Gaudy, A.F. Jr., and Martin, J. (1991). Application of a Process Control Model Calibrated with Respirometry to Predict Full-Scale Activated Sludge Performance. *Proceedings, 45th Purdue Industrial Waste Conference,* Lewis Publishers, Chelsea, Michigan, pp. 501–508.

Colvin, R.J., Rozich, A.F., Hough, B.J., and Gaudy, A.F. Jr. (1991). "Use of Respirometry to Evaluate Biodegradability of Emulsified Specialty Chemical Products" (with R.J. Colvin, B. Hough, and A.F.

Gaudy, Jr.) *Proceedings, 45th Purdue Industrial Waste Conference,* Lewis Publishers, Chelsea, Michigan, pp. 477–486.

Constable, S.W., Rozich, A.F., DeHaas, R., and Colvin, R.J. (1991). "Respirometric Investigation of Activated Sludge Bioinhibition by Cobalt/Manganese Catalyst," *Presented, 46th Annual Industrial Waste Conference,* Purdue University, West Lafayette, IN, May, 1991.

Gaudy, A.F. Jr., Rozich, A.F., and Lowe, W.L. (1987). "Design Criteria for Treatment of Combined Waste," *Presented, Annual Conference of the Environ. Eng. Div., ASCE,* Orlando, FL.

Peil, K.M., and Gaudy, A.F. Jr. (1971). "Kinetic Constants for Aerobic Growth of Microbial Populations Selected with Various Single Compounds and with Municipal Wastes as Substrates." *Appl. Microbiol., 21,* pp. 253–256.

Rozich, A.F., and Colvin, R.J. (1990). "Formulating Strategies for Activated Sludge Systems," *Water Engineering & Management, 137,* 10, pp. 39–41.

Rozich, A.F., Usinowicz, P.J., and Colvin, R.J. (1991). Operation of a Superfund Site Biological Wastewater Treatment Plant Using Predictive Respirometric Analyses. *Presented, 64th Annual WPCF Conference,* Toronto, Canada.

Rozich and Gaudy, Inc., (1986). "Development of a Process Control Plan for Managing Waste Inhibition at the Patapsco Wastewater Treatment Plant," Engineering Report to the City of Baltimore, MD, Wastewater Facilities Division.

Rozich and Gaudy, Inc., (1986). "Determination of the Numerical Values of the Biokinetic Constants and Implications to the Design of the Expanded Facilities for the Patapsco Wastewater Treatment Plant," Engineering Report to the City of Baltimore, MD, Wastewater Facilities Division.

Appendix: Computer Programs

PART I: LISTING OF THE COMPUTER PROGRAM USED TO FIT GROWTH DATA TO THE MONOD FUNCTION TO DETERMINE THE BIOKINETIC CONSTANTS μ_{max} AND K_s

THIS PROGRAM UTILIZES THE CURVE FITTING ROUTINE "DUNLSF" FOR FITTING BATCH GROWTH DATA TO THE MONOD FUNCTION.

```
        IMPLICIT REAL *8 (A-H, O-Z)
        EXTERNAL PARAB
        INTEGER M,N,LDFJAC,IPARAM(6),NSIG
        DIMENSION XGUESS(2),XSCALE(2), FSCALE(100)
        DIMENSION RPARAM(7), X(2), FVEC(100), FJAC(100,2),
        F(100)
        COMMON /ULF/Y(100),V(100)
        DATA FSCALE/100*1.0D0/,XSCALE/2*1.0D0/
        N = 2
140     FORMAT(/ ' ENTER NUMBER OF DATA POINTS, I.')
150     FORMAT (/ ' ENTER, THE EXPERIMENTAL S0 AND U
        DATA, PLEASE.')
155     FORMAT (/ ' ONE DATA SET PER LINE PRINTING S0
        FIRST.')
160     FORMAT (/ ' ENTER INITIAL GUESSES FOR PARAMETERS
        "UMAX" AND "KS" 1, PLEASE/')
162     FORMAT (/ ' THE INITIAL GUESS FOR UMAX IS YOUR
        LARGEST EXPERIMENTAL GROWTH RATE (U).')
164     FORMAT (/ ' THE INITIAL GUESS FOR KS IS YOUR
        SMALLEST INITIAL SUBSTRATE CONCENTRATION
        (SO).')
165     FORMAT (/ ' ENTER VALUES FOR EPS AND NSIG (START
        WITH EPS = 0.0 AND NSIG = 3).')
170     FORMAT (/5X,' UMAX (1/H) = ',F10.3)
173     FOPRMAT (/5X,' KS (MG COD/L) = ',F10.1)
```

```
175   FORMAT (/5X,' SSQ  =',F10.6)
180   FORMAT (/' DO YOU WISH TO CHANGE YOUR INITIAL
      GUESSES? YES = 1 AND NO = 2')
185   FORMAT (/' DO YOU WISH TO TRY ANOTHER SET OF
      DATA? YES = 1 AND NO = 2')
190   FORMAT (/' NICE TALKING TO YOU.')
 10   WRITE (6,140)
      READ(5,*) M
      LDFJAC-100
      WRITE(6,150)
      WRITE(6,155)
      DO 15,I = 1,M
 15   READ(5,*) V(I),Y(I)
 17   CONTINUE
      WRITE(6,160)
      WRITE(6,162)
      WRITE(6,164)
      READ(5,*) XGUESS(1),XGUESS(2)
      CALL DU4LSF(IPARAM,RPARAM)
      WRITE(6,165)
      READ(5,*) EPS,NSIG
      RPARAM(4) = EPS
      IPARAM(2) = NSIG
      IPARAM(3) = 1000
      IPARAM(4) = 1000
      CALL DUNLSF FPARAB,M,N,XGUESS,XSCALE,
      FSCALE,IPARAM,RPARAM,X,FVEC,FJAC,LDFJAC)
      WRITE (6,170) X(1)
      WRITE (6,173) X(2)
      DO 20, K = 1,M
      SSQ = FVEC(K)**2 + SSQ
 20   CONTINUE
      WRITE(6,175) SSQ
      WROTE(6,180)
      READ(5,*) ANS
      IF (ANS.EQ.1) GO TO 17
      WRITE(6,185)
      READ(5,*) ANS2
      IF (ANS2.EQ.1) GO TO 10
      WRITE(6,190)
      STOP
```

```
      END
      SUBROUTINE PARAB(M,N,X,F)
      IMPLICIT REAL*8 (A-H,O-Z)
      INTEGER M,N
      DIMENSION X(2), F(100)
      COMMON /UFL/Y(100),V(100)
      DO 5, I = 1,M
    5 F(I) = Y(I) - (X(1)*V(I))/(X(2) + V(I))
      RETURN
      END
```

PART II: LISTING OF THE COMPUTER PROGRAM USED TO FIT GROWTH DATA TO THE HALDANE FUNCTION TO DETERMINE THE BIOKINETIC CONSTANTS μ_{max}, K_s AND K_i

"NONLIN FORTRAN" CURVE FITTING PROGRAM
FOT FITTING BATCH GROWTH DATA TO
THE HALDANE GROWTH FUNCTION

```
      IMPLICIT REAL*8 (A-H,O-Z)
      EXTERNAL FUNC1
      INTEGER N,M,IPARAM(6),LDFJAC
      DIMENSION XGUESS(3),XSCALE(3),FSCALE(100),
      RPARAM(7)
      DIMENSION X(3),FVEC(100),FJAC(100,3),F(100)
      COMMON/BLOCK1/XEXPT(100),YEXPT(100)
      DATA FSCALE/100*1.0D0/,XSCALE/3*1.0d0/
      DATA IN,IOUT /5,5/
      CALL ERSET(4,-1,0)
    5 WRITE(6,10)
   10 FORMAT(' ENTER THE NUMBER OF DATA POINTS,
      PLEASE.')
      READ(5,*) M
      WRITE(6,15)
   15 FORMAT(' ENTER THE LARGEST EXPERIMENTAL
      GROWTH RATE, PLEASE')
      READ(5,*) USTAR
      WRITE(6,20)
```

```
20    FORMAT(' ENTER THE EXPERIMENTAL SO AND U DATA,
      PLEASE.')
      WRITE (6,25)
25    FORMAT(' ONE SET OF DATA PER LINE STARTING WITH
      SO.')
      DO 30 I = 1,M
30    READ(5,*) XEXPT(I),YEXPT(I)
32    WRITE(6,35)
35    FORMAT(' ENTER A VALUE FOR "EPS" (0.1 TO 0.00000001),
      PLEASE')
      READ(5,*) EPS
      N = 3
      LDFJAC = 100
      SSQ = 0.0
      INPUT GUESS MATRIX:  USTAR < UMAX < 10*USTAR
                             1.0  <  KS  <   300.0
                             1.0  <  KI  <  1500.0
      UMAXH = 10*USTAR
      UMAXL = USTAR
      DO 500 I = 1,8
      XGUESS(1) = UMAXL*(1.28571438*FLOAT(I)-0.28571438)
      DO 600 J = 1,12
      XGUESS(2) = 27.181818*FLOAT(J)-26.181818
      DO 700 K = 1,15
      XGUESS(3) = 107.071429*FLOAT(K)-106.071429
      CALL DU4LSF(IPARAM,RPARAM)
      IPARAM(1) = 1
      IPARAM(2) = 3
      IPARAM(3) = 20000
      IPARAM(4) = 20000
      RPARAM(1) = 0.0
      RPARAM(4) = EPS
      CALL DUNLSF(FUNC1,M,N,XGUESS,SCALE,FSCALE,
      IPARAM,RPARAM,X,FVEC,FJAC,LDFJAC)
      IF((IPARAM(3).EQ.20000).OR.(IPARAM(4).EQ.20000)) GOTO
      79
      IF((X(1).GT.UMAXL).AND.(X(1).LT.UMAXH)) GOTO 40
      GOTO 700
40    IF((X(2).GT.1).AND.(X(2).LT.300)) GOTO 50
      GOTO 700
50    IF((X(3).GT.1).AND.(X(3).LT.1500)) GOTO 60
```

```
700    CONTINUE
600    CONTINUE
500    CONTINUE
 79    WRITE(6,80)
 80    FORMAT(//,' CURVE FIT HAS FAILED! TRY AGAIN.')
       GOTO 82
 60    DO 70 L = 1,M
       SSQ = FVEC(L)**2 + SSQ
 70    CONTINUE
       GOTO 90
 82    WRITE(6,85)
 85    FORMAT(' WOULD YOU LIKE TO TRY A NEW "EPS"
       VALUE? YES = 1 AND NO = 2')
       READ(5,*) ANS
       IF(ANS .EQ. 1) GOTO 32
       GOTO 155
 90    WRITE(IOUT,100) X(1)
100    FORMAT(/5X,' UMAX (1/HOURS) = ',F10.4)
       WRITE(IOUT,110) X(2)
110    FORMAT(/5X,' KS (MG COD/L) = ',F10.4)
       WRITE(IOUT,120) X(3)
120    FORMAT(/5X,' KI (MG COD/L) = ',F10.4)
       WRITE(IOUT,130) SSQ
130    FORMAT(/5X, SSQ  = ',F10.5)
       GOTO 82
155    WRITE(6,160)
160    FORMAT(' WOULD YOU LIKE TO TRY ANOTHER SET OF
       DATA? YES = 1 AND NO = 2')
       READ(5,*) ANS2
       IF(ANS2 .EQ. 1) GOTO 5
       STOP
       END
       SUBROUTINE FUNC1 (M,N,X,F)
       IMPLICIT REAL*8 (A-H,O-Z)
       DIMENSION F(100),X(3)
       COMMON/BLOCK1/XEXPT(100),YEXPT(100)
       DO 800 I = 1,M
       F(I) = YEXPT(I)-(X(1)/(1 + (X(2)/XEXPT(I)) + (XEXPT(I)/
       X(3))))
800    CONTINUE
       RETURN
```

END

Bounding Rules for "Nonlin Fortran"

Coefficient	Allowable Range
μ_{max}	$\mu^* \leq \mu_{max} \leq 10\,\mu^*$
K_s	$1 \leq K_s \leq 300$
K_i	$1 \leq K_i \leq 1500$

INDEX